PRODUCT DESIGN 4

PRODUCT DESIGN 4

Compiled and Edited by Joel Sokolov

The Library
of Applied
Design

An Imprint of **PBC** INTERNATIONAL, Inc.

Distributor to the book trade in the United States and Canada:

Rizzoli International Publications Inc.
300 Park Avenue South
New York, NY 10010

Distributor to the art trade in the United States:

Letraset USA
40 Eisenhower Drive
Paramus, NJ 07652

Distributor to the art trade in Canada:

Letraset Canada Limited
555 Alden Road
Markham, Ontario L3R 3L5, Canada

Distributed throughout the rest of the world:

Hearst Books International
105 Madison Avenue
New York, NY 10016

Library of Congress Cataloging-in-Publication Data
Sokolov, Joel.
    Product design 4/compiled and edited by Joel Sokolov.
        p.   cm.
    ISBN 0-86636-129-4
    1. Design, Industrial.   I. Title.   II. Title: Product design four.
TS171.S59   1990   90-7030
745.2—dc20   CIP

*CAVEAT—Information in this text is believed accurate, and
will pose no problem for the student or casual reader.
However, the authors were often constrained by information
contained in signed release forms, information that could
have been in error or not included at all. Any
misinformation (or lack of information) is the result of failure
in these attestations. The authors have done whatever is
possible to insure accuracy.*

Color separation, printing and binding by
Toppan Printing Co. (H.K.) Ltd. Hong Kong

Typography by
Toledo Typesetting

10 9 8 7 6 5 4 3 2 1

ACKNOWLEDGEMENTS

Thanks to *Leo Blackman, Lois Lambert, Don Ruddy, Tim Veness, Iain Sinclair, Tucker Viemeister, Jim Pagella, Gordon Perry and Helen Shirk for being part of PRODUCT DESIGN 4.*

Extra special thanks to *Kevin Clark for having the confidence in me and the opportunity to work on PRODUCT DESIGN 4 with him. Extra special thanks to John Wuchte for being himself.*

Special thanks to *Erika Arroyo, Art et Industrie, Ken Barber, Drew & Catherine Betterton, Carlo Bonoan, Howard Burchman, Joanne DiLorenzo, Joseph Dionisio, Dorland Mountain Arts Colony, Ria Eagan, The Gallery of Functional Art, Bob Gereke, Audrey Goldstein, Judy Herzl, Peter Hoyle, Johanna Immerman, Innovative Design Fund, Jim Lambert, Carlo Lubrano, Craig Lyman, Frank Marx, Trina Marx, Fran Preisman, Dominique Rageneau, Mark Robbins, Bob Smith, All the Sokolovs and the Miller, David Steinberg, Caroline Stern, Lisa Wolpe, and Yvette and Face, all of whom in their own ways contributed to PRODUCT DESIGN 4.*

CONTENTS

INTRODUCTION

It's hard to imagine that for every one of the multitude of products we encounter, there is a product designer somewhere out there. Using the materials in their environments, 'primitive' designers saw their needs then created objects to make their lives easier and more productive. While this impetus still drives designers in their endeavors, it is a far cry from one aspect of the present state of product design. The era of products for product's sake – with the bottom line being whether the product can be sold – has arrived, and is still with us. On the other hand, innumerable product designers are creating innumerable products for a myriad of uses, old and new, and are diligently answering the needs of our daily lives with a positive view of the future. As we enter the '90s, the plurality of approaches to product design promises a lot of pleasant surprises. It's just those surprises, which inspire us and open the way for new visions and new possibilities, that shape where we are going.

Product design is an international language. The challenge for product designers in the '90s is to take more responsibility for the products we design and their ultimate impact. We know we can design and build products intended only to be sold and trashed, as well as too many very destructive products. The '90s call for products with a social conscience. Whether a designer creates a one-of-a-kind product intended only to uplift the spirit, an everyday utilitarian object, or a very specific piece of medical machinery, attention must be paid to the way a product interacts and integrates with a greater environment. Product designers have the power to change the way people think, because products can open new perspectives on the world around us. The products we use and surround ourselves with reflect us and comment on our customs and values. We are the products we use. We are advertisements for the products we use, just as they advertise us (". . . If you don't look good, we don't look good. . ."). Whole generations have grown up and taken for granted products that were unimaginable in the recent past. As we continue to create new products and new needs for them, product designers will further explore the blending of technology, art and craft. Product design, like art, has the power to influence. Just as 'conversational' art can make us examine our values personally and as a culture, a product can have the same effect. Ordinary (and extraordinary) products have value and influence beyond their practical functions. This might be a simply latch key;

a solar energy panel that produces all necessary power without pollution, for free; a chair that makes you think about sitting a little differently; a cure for AIDS; or an easily programmable VCR. A well designed product can mean a lot.

PRODUCT DESIGN 4 represents only a small cross-section of the numerous and talented product designers out there. From every swizzle stick to sneakerphone to skyscraper, product designers are doing their best to fulfill our practical needs, while at the same time, surprise and inspire us with the inventiveness, quality and craftsmanship of their work.

1 Tabletop

CONSCIOUSLY OR UNCONSCIOUSLY, we choose products that touch us emotionally for some reason; those which reveal a sense of life or humor, tranquillity or order, risk or unpredictability; those that help us make contact with what we value or miss most in ourselves, and in our lives.

The vessel provides an intimate format for me, partly due to the role it has played in man's history, but also because it allows a private conversation between me and the object. Patterns of growth and images I observe in nature become metaphors for what concerns me on a personal level: the fragility of new growth, displaying its resiliency and strength, by existing beside fragments of nature past its prime; the perception of violence and beauty in the inevitable disintegration; the tenuous structure, resulting from age and stress, tenaciously surviving despite the withdrawal of nourishment. The symmetrical format provides an element of stability for me, an instinctively understood point from which I start my journey.

In the midst of a continuous effort to streamline the objects we use and the way we live with them, there exists a counterpoint: the seemingly ungainly, unpretentious object, which reminds us of the beauty of the odd man out.

Helen Shirk is a professor of Art at San Diego State University, but her design expertise has been recognized on a much wider scale.

Besides two Fellowship grants to the National Endowment for the Arts, her impressive list of exhibitions includes the Carnegie Museum of Art, Pittsburgh; the National Museum of Modern Art, Japan; the Schmuckmuseum, West Germany; the Smithsonian Institution, Washington, D.C.; the American Craft Museum, New York; the Oxford Gallery, England; and the Musée Des Arts Decoratifs, Paris.

Ms. Shirk's designs have also graced the pages of the International Design Yearbook 4; Contemporary American Craft Art: A Collector's Guide; Craft Today: Poetry of the Physical; Kunst und Handwerk; Women Artists News; Metropolis; the San Diego Union and San Diego Home and Garden. She attended Kunsthaandvaerkerskolen in Copenhagen, Denmark, and received an M.F.A. from Indiana University and a B.S. from Skidmore College.

HELEN SHIRK
Designer

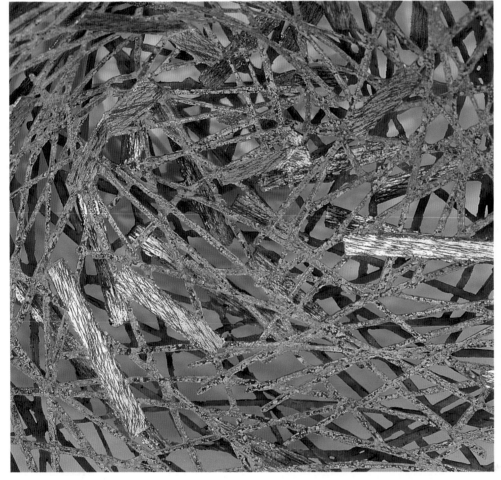

PRODUCT
"Intimate Partners"
DESIGNER
Helen Shirk
DESCRIPTION
Spray-etched double bowls
PHOTOGRAPHY
Will Gullette

PRODUCT
"On the Edge—Boundary"
DESIGNER
Helen Shirk
DESCRIPTION
Gold-leafed bowl
PHOTOGRAPHY
Will Gullette

PRODUCT
"On the Edge—Memory"
DESIGNER
Helen Shirk
DESCRIPTION
Center of bowl is spray-etched, patinated and then cut back to expose some of the underlying copper.
PHOTOGRAPHY
Will Gullette

PRODUCT
"Limelight" Plate
DESIGNER
Göran Wärff
FIRM
Kosta Boda
PHOTOGRAPHY
Sten Robert

PRODUCT
Arturo Tray
DESIGNER
Roberto Marcatti
FIRM
Lavori in Corso
MANUFACTURER
Lavori in Corso
DESCRIPTION
Tray shaped with a laser; copper
rivets and brushed finish
PHOTOGRAPHY
Foto Mosna

PRODUCT
Y.M.D. IMONO Platter
DESIGNER
Takenobu Igarashi
FIRM
Igarashi Studio
MANUFACTURER
Yamasho Casting
DESCRIPTION
Created by using traditional cast
iron methods used in Yamagata
Prefecture for over 900 years
PHOTOGRAPHY
Masaru Mera
DISTRIBUTOR
Yamada Shomei Lighting Co.,
Y.M.D. Division

PRODUCT
Petro Dinnerware
DESIGNER
Ann Morhauser
FIRM
Annieglass Studio
PHOTOGRAPHY
Viktor Budnik

PRODUCT
"XUM" 5-piece Place Setting
DESIGNER
Robert Wilhite
FIRM
Bissell & Wilhite
MANUFACTURER
Bissell & Wilhite

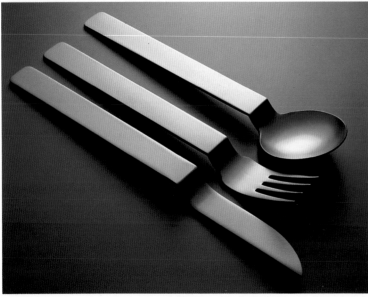

PRODUCT
Com Cutlery
DESIGNER
Makoto Komatsu
FIRM
Eastern Accent International Inc.
DESCRIPTION
Sandblasted stainless flatware.
DISTRIBUTOR
Eastern Accent International Inc.

PRODUCT
Y.M.D. Stainless Flatware
DESIGNER
Takenobu Igarashi
FIRM
Igarashi Studio
MANUFACTURER
Tsubame Shinko Industrial Co., Ltd.
DESCRIPTION
Series from flatware to desktop accessories using the famed craftsmanship of the metalworkers in Tsubame City, Niigata Prefecture
PHOTOGRAPHY
Masaru Mera
DISTRIBUTOR
Yamada Shomei Lighting Co., Y.M.D. Division

PRODUCT
"EXL" 5-piece Place Setting
DESIGNER
Robert Wilhite
FIRM
Bissell & Wilhite
MANUFACTURER
Bissell & Wilhite

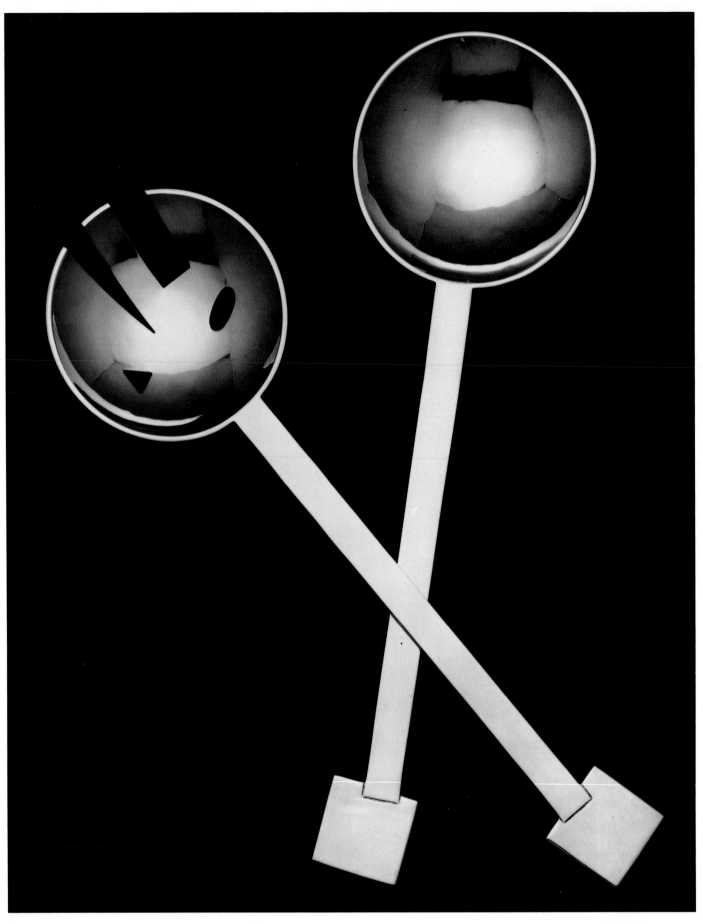

PRODUCT
"XUM" Salad Service
DESIGNER
Robert Wilhite
FIRM
Bissell & Wilhite
MANUFACTURER
Bissell & Wilhite

PRODUCT
Carrot Knife, Tomato Knife, Bread
Knife and Chicken Leg Slicer
DESIGNER
Michael Schneider
MANUFACTURER
Giesser
CLIENT
Markuse Corp.

PRODUCT
Utility and Paring Knives
DESIGNER
Wolfgang Geisler
MANUFACTURER
Giesser
CLIENT
Markuse Corp.

PRODUCT
Carving Set
DESIGNER
Justus Theinert
MANUFACTURER
Giesser
DESCRIPTION
Silverplated and ebony wood-
handled
CLIENT
Markuse Corp.

PRODUCT
Sunny
DESIGNER
Monica Backström
FIRM
Kosta Boda AB
PHOTOGRAPHY
Rolf Morlin

PRODUCT
Athena
DESIGNER
Monica Backström
FIRM
Kosta Boda AB
PHOTOGRAPHY
Rolf Morlin

PRODUCT
Birdy
DESIGNER
Ulrica Hydman-Vallien
FIRM
Kosta Boda AB
MANUFACTURER
Kosta Boda AB
PHOTOGRAPHY
Janne Bengtsson

PRODUCT
"Caramba" Jars and Bowl
DESIGNER
Ulrica Hydman-Vallien
FIRM
Kosta Boda AB
MANUFACTURER
Orrefors

PRODUCT
"Open Minds"
DESIGNER
Ulrica Hydman-Vallien
FIRM
Kosta Boda AB
MANUFACTURER
Orrefors

PRODUCT
Space
DESIGNER
Monica Backström
FIRM
Kosta Boda AB
PHOTOGRAPHY
Rolf Mörlin

PRODUCT
Helix
DESIGNER
Ann P. Wahlström
FIRM
Kosta Boda
MANUFACTURER
Kosta Boda AB
DESCRIPTION
Full lead crystal; blown and cast
glass, gray underlay color, cut and
polished
PHOTOGRAPHY
Sten Robert

PRODUCT
Knossos I
DESIGNER
Ann P. Wahlström
FIRM
Kosta Boda
MANUFACTURER
Kosta Boda AB
DESCRIPTION
Full lead crystal; blown and cast
glass, violet underlay color, cut
and polished
PHOTOGRAPHY
Sten Robert

PRODUCT
Falstabo Amber I
DESIGNER
Matz Borgstrom
FIRM
Kosta Boda AB
MANUFACTURER
Orrefors
DESCRIPTION
Abstract glass bowl

PRODUCT
"Modulus Vivande" — Boggie,
Junction and Voltage
DESIGNER
Gunnel Sahlin
FIRM
Kosta Boda
MANUFACTURER
Kosta Boda AB
DESCRIPTION
Full lead crystal, assembled hot
PHOTOGRAPHY
Jan Bengtsson

PRODUCT
Circus
DESIGNER
Ann P. Wahlström
FIRM
Kosta Boda
MANUFACTURER
Kosta Boda AB
DESCRIPTION
Turquoise and black transparent
full lead crystal; Graal technique
PHOTOGRAPHY
Sten Robert

PRODUCT
"Modulus Vivande" — Out of the
House, U-Turn
DESIGNER
Gunnel Sahlin
FIRM
Kosta Boda
MANUFACTURER
Kosta Boda AB
DESCRIPTION
Full lead crystal, assembled hot
PHOTOGRAPHY
Jan Bengtsson

PRODUCT
Circus
DESIGNER
Ann P. Wahlström
FIRM
Kosta Boda
MANUFACTURER
Kosta Boda AB
DESCRIPTION
Red and black opaque full lead
crystal; Graal technique
PHOTOGRAPHY
Sten Robert

PRODUCT
Venus
DESIGNER
Ann P. Wahlström
FIRM
Kosta Boda
MANUFACTURER
Kosta Boda AB
DESCRIPTION
Full lead crystal; blown glass with
cast details, blue and black
underlay color
PHOTOGRAPHY
Sten Robert

PRODUCT
Movement
DESIGNER
Göran Wärff
FIRM
Kosta Boda

PRODUCT
Cleopatra
DESIGNER
Ulrica Hydman-Vallien
FIRM
Kosta Boda AB, Sweden
MANUFACTURER
Kosta Boda AB
DESCRIPTION
Hand-blown crystal
PHOTOGRAPHY
Janne Bengtsson

PRODUCT
Inevra & Artu
DESIGNER
Roberto Marcatti
FIRM
Nuova Vilca s.a.s.
MANUFACTURER
Lavori in Corso
DESCRIPTION
Crystal glasses in steel support
PHOTOGRAPHY
Foto Mosna

PRODUCT
Mambo
DESIGNER
Gunnel Sahlin
FIRM
Kosta Boda
MANUFACTURER
Kosta Boda AB
DESCRIPTION
Lead crystal with a color underlay
PHOTOGRAPHY
Sten Robert

PRODUCT
Latin Love
DESIGNER
Gunnel Sahlin
FIRM
Kosta Boda
MANUFACTURER
Kosta Boda AB
DESCRIPTION
Lead crystal with a color underlay
PHOTOGRAPHY
Sten Robert

PRODUCT
"Future Form" Tea or Coffee
Service (glossy finish)
DESIGNER
Marek Cecula
MANUFACTURER
Marek Cecula
DESCRIPTION
Set is composed of 4 cups and
1 pot
CLIENT
Contemporary Porcelain
PHOTOGRAPHY
Bill Waltzer

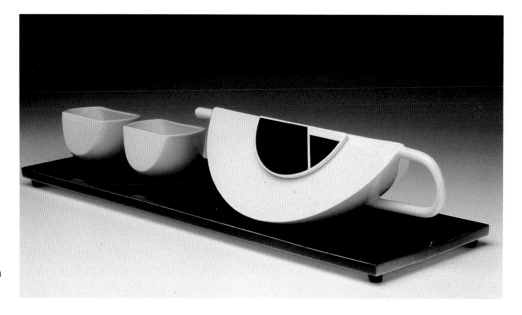

PRODUCT
Ceremonial Set II
DESIGNER
Marek Cecula
MANUFACTURER
Marek Cecula
DESCRIPTION
For either tea or sake
CLIENT
Contemporary Porcelain
PHOTOGRAPHY
Bill Waltzer

PRODUCT
"Graphic Set" Tea or Coffee
Service
DESIGNER
Marek Cecula
MANUFACTURER
Marek Cecula
DESCRIPTION
The set is designed with a
3-dimensional content where
color and form become an
integrated part of the whole
CLIENT
Contemporary Porcelain
PHOTOGRAPHY
Bill Waltzer

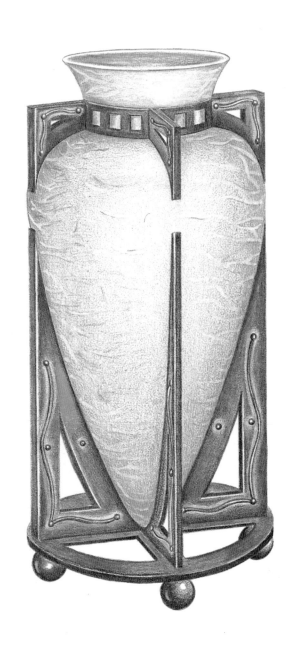

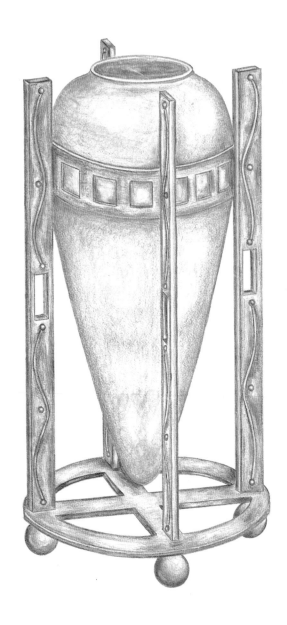

PRODUCT
Vases
DESIGNER
Joseph DiGiacomo
DESCRIPTION
Conceptual sketches of architecturally-inspired vases: at left, a copper finish base with frosted crackle glass; at right, a pewter finish base with frosted glass.

PRODUCT
Source/Fold Space Square
DESIGNER
Matz Borgstrom
FIRM
Kosta Boda AB
MANUFACTURER
Orrefors
DESCRIPTION
Abstract glass sculptures

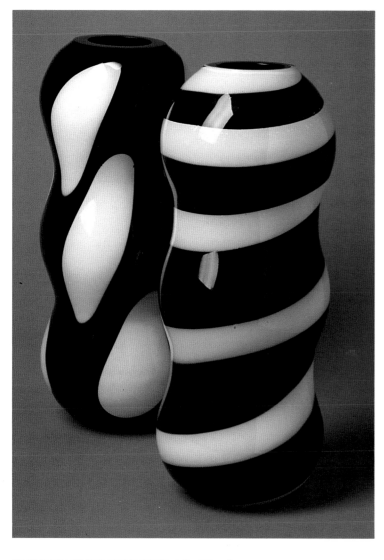

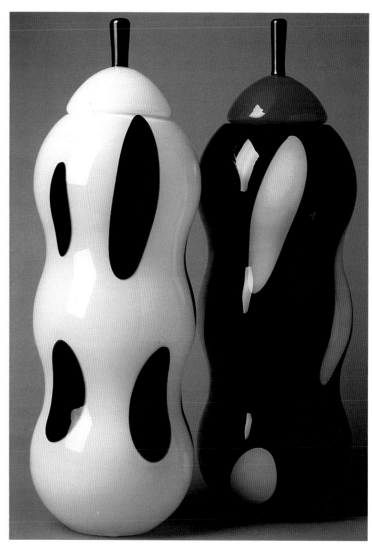

PRODUCT
Puppis
DESIGNER
Gunnel Sahlin
FIRM
Kosta Boda
MANUFACTURER
Kosta Boda AB
DESCRIPTION
Lead crystal vase
PHOTOGRAPHY
Katarina Rothfjäll

PRODUCT
Cosmic Embrace
DESIGNER
Göran Wärff
FIRM
Kosta Boda

31

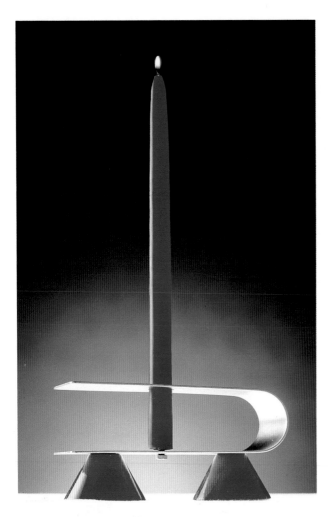

PRODUCT
Candlesticks
DESIGNERS
Milton Glaser, Aldo Cibia,
Aldaberto Pironi
MANUFACTURER
Alessi
CLIENT
Markuse Corp.

PRODUCT
Pippo
DESIGNER
Roberto Marcatti
FIRM
Lavori in Corso
MANUFACTURER
Lavori in Corso
DESCRIPTION
Candle holder in folding sheet
iron
PHOTOGRAPHY
Foto Mosna

PRODUCT
Candelabra
DESIGNER
Arman
FIRM
A/D
DESCRIPTION
To be cast in silver gilt
PHOTOGRAPHY
Ken Schles
DISTRIBUTOR
A/D

PRODUCT
IPSOS Vase
DESIGNER
John Beckmann; Antonia Polizzi
FIRM
Prologue 2000
DISTRIBUTOR
Prologue 2000

PRODUCT
Y.M.D. IMONO Candlestand
DESIGNER
Takenobu Igarashi
FIRM
Igarashi Studio
MANUFACTURER
Yamasho Casting
DESCRIPTION
Created by using traditional cast iron methods used in Yamagata Prefecture for over 900 years
PHOTOGRAPHY
Masaru Mera
DISTRIBUTOR
Yamada Shomei Lighting Co., Y.M.D. Division

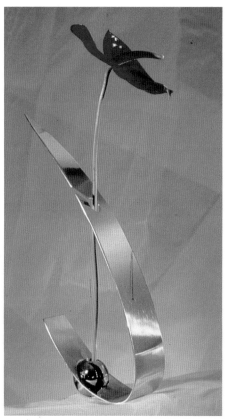

PRODUCT
"Herbie" Fruit Dish
DESIGNER
Luciano Deviá
FIRM
Luciano Deviá Design Associados
CLIENT
Artgramar Ltda.

PRODUCT
Cheops Mirror, X-19 Fotoframe,
Ursa Major Tray
DESIGNER
Lorelei Hamm
DESCRIPTION
Wire glass on Ursa Major Tray
allows for extreme heat to be
placed on it
PHOTOGRAPHY
Bill de Michele

PRODUCT
Stainless Steel Cocktail Shaker
(1925)
DESIGNER
Marianne Brandt
MANUFACTURER
Alessi
CLIENT
Markuse Corp.
PHOTOGRAPHY
Aldo Ballo

PRODUCT
Campari Cocktail Shaker
DESIGNER
Matteo Thun
MANUFACTURER
Alessi
CLIENT
Markuse Corp.

PRODUCT
Colander
DESIGNER
Philippe Starck
MANUFACTURER
Alessi
DESCRIPTION
Stainless steel with brass legs
CLIENT
Markuse Corp.

PRODUCT
Cooking Pots — Vikings
DESIGNER
Heiie Damkjaer
PHOTOGRAPHY
Milano Anthrazit

PRODUCT
Coffee Mill
DESIGNER
Riccardo Dalisi
MANUFACTURER
Alessi
CLIENT
Markuse Corp.

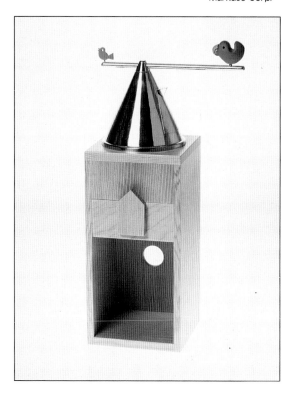

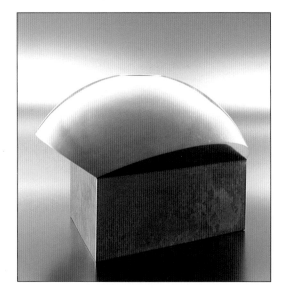

PRODUCT
"EN" Wall Vase
DESIGNER
Masayuki Kurokawa
FIRM
Eastern Accent International Inc.
DESCRIPTION
Uses the traditional coloring method of Takaoka to achieve the *ni-iro* (brown/dark brown color)
DISTRIBUTOR
Eastern Accent International Inc.

PRODUCT
Children's Clocks
FIRM
Canetti Design Group
MANUFACTURER
Canetti Inc.
DESCRIPTION
Solid wood with some removable pieces
CLIENT
Canetti Inc.

PRODUCT
Dome wall clock
FIRM
Canetti Design Group
MANUFACTURER
Canetti Inc.
DESCRIPTION
Clock of metal sheet with curved hands
CLIENT
Canetti Inc
PHOTOGRAPHY
Color Track

PRODUCT
Timers
DESIGNER
Roberto Pezzetta
MANUFACTURER
Wiki Due
DESCRIPTION
0–60 minute spring-loaded timer
CLIENT
Markuse Corp.

PRODUCT
Asteroid wall clocks
FIRM
Canetti Design Group
DESCRIPTION
Glass and metal with second
hand continuous rotation
CLIENT
Canetti Inc.
PHOTOGRAPHY
Color Track

PRODUCT
Spring Collection wall clocks
FIRM
Canetti Design Corp.
MANUFACTURER
Collectors Collection
DESCRIPTION
Made of rubber frame, glass and
plastic back.
CLIENT
Canetti Inc.
PHOTOGRAPHY
Color Track

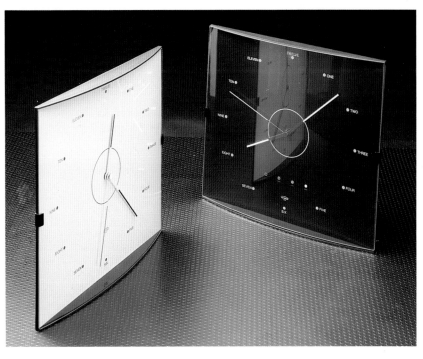

2 Textiles

IN SOME CASES, PEOPLE ARE THE sum total of the products they use. There is no doubt we live in the age of consumerism, with society offering us constant access to a vast array of products for any particular need. In general, I feel we are not defined by the products we choose, as much as we are reflected by them. We use varying criteria: practical, aesthetic, financial, political, ecological – even for the most mundane of choices. I think there is a trend toward more personal self-expression; our products closely reflect our tastes, as well as a growing concern for the environment and how products and their use impact on it. Choices are no longer defined by what may be the easiest or most practical, but by what may be the most interesting or environ-

mentally safe. The products we use reflect who we are, economically, aesthetically and politically.

My design philosophy lies in a fascination with complexity and intricacy. I like to design textiles that at first glance appear simple and straight-forward, but upon closer examination, reveal a multitude of color, texture and pattern. It is that juxtaposition of the obvious and the subtle, the visible and the invisible, that inspires my work.

It is important in my work to include elements that at least on a subtle level, retain the integrity of something hand-made. I feel strongly that the role of a craftsperson is a vital one, making products available that reflect the presence of the hand that made them. By doing so, something unique is created – by myself as the designer, and by the person who chooses my work.

As for design trends in the '90s and beyond, I would say the theme is anything goes. We continue to see historically that trends repeat themselves. What was popular in the '50s or '60s may be popular again now, but these trends are often recycled with some new twist. More and more, styles from the past are mixed with things that are new and contemporary. I see an increased use of design in more personal ways, people making their own particular statement, instead of relying upon a trend or style in its original state. There is also a greater concern for the environment, by both the designer as well as the consumer. That will result in more natural and relaxed trends.

TIM VENESS
Textile Designer

Tim Veness is an award-winning designer, based in San Francisco, who specializes in textiles and rugs. His work has been seen in publications such as Harper's Bazaar, Gentleman's Quarterly, Interview, Esquire and Women's Wear Daily.

In 1984, he founded VENESS, marketing a line of men's sweaters and home furnishings fabrics. He has been honored with the "Coty American Fashion Critics' Special Menswear Award," as well as nominations for the Cutty Sark Men's Fashion Award for "Most Promising American Designer," and the "Focus Magazine / Golden Shears Award."

Mr. Veness attended the School of Art & Design at the Rochester Institute of Technology, San Francisco City College, and San Francisco State University.

PRODUCT
"Andover" wall covering
FIRM
Richard E. Thibaut
MANUFACTURER
Richard E. Thibaut, Inc.

PRODUCT
"Silence" wall covering
MANUFACTURER
J.M. Lynne Co., Inc.
CLIENT
J.M. Lynne Co., Inc.

PRODUCT
Provénce Collection (Lacrosse)
DESIGNER
Laura Deubler Mercurio
MANUFACTURER
Adam James Textiles
CLIENT
Adam James Textiles

PRODUCT
Design Latino Carpet
DESIGNER
Davide Mercatali
MANUFACTURER
AB2
DESCRIPTION
Pure wool

PRODUCT
Rugs
DESIGNER
Tim Veness
DESCRIPTION
Handwoven
PHOTOGRAPHY
Christine Alicino

PRODUCT
Bastille
DESIGNER
Andree Putman
CLIENT
Stendig Textiles
PHOTOGRAPHY
Bill Whitehurst Studios
DISTRIBUTOR
Stendig Textiles

PRODUCT
"Vortex" Rug
DESIGNERS
Carolyn and Vincent Carleton
FIRM
Carleton Designs
MANUFACTURER
Carleton Designs
DESCRIPTION
Handwoven, hand-dyed, inversible rugs
PHOTOGRAPHY
Dick Boehme

PRODUCT
Regina
DESIGNER
Laura Deubler Mercurio
MANUFACTURER
Adam James Textiles
CLIENT
Adam James Textiles

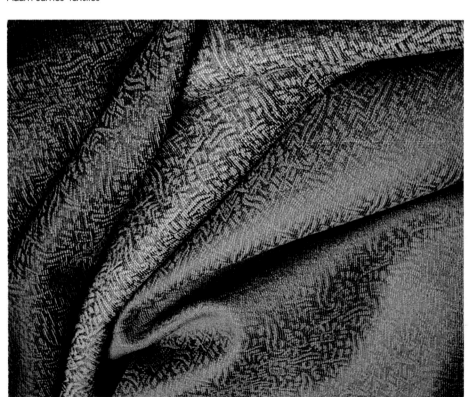

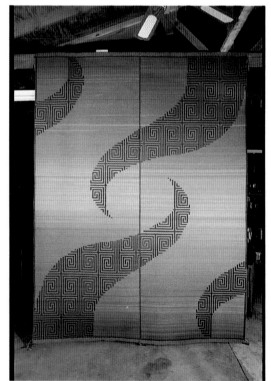

PRODUCT
"Flamestitch" Rug (detail)
DESIGNER
Carolyn and Vincent Carleton
FIRM
Carleton Designs
MANUFACTURER
Carleton Designs
DESCRIPTION
Hand Jacquard woven, hand-dyed, inversible rug
CLIENT
Larry Hahn
PHOTOGRAPHY
Sean Sprague
AWARD
'Best Contemporay Rug Design', ROSCOE Product Design Award, 1989

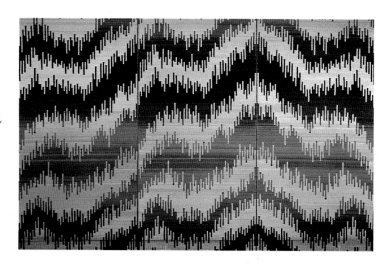

PRODUCT
'Saf' Rug (front view)
DESIGNER
Carolyn and Vincent Carleton
FIRM
Carleton Designs
MANUFACTURER
Carleton Designs
DESCRIPTION
Hand woven, hand-dyed, inversible rugs
CLIENT
Jane and Robert Socolow
PHOTOGRAPHY
Dennis Geaney

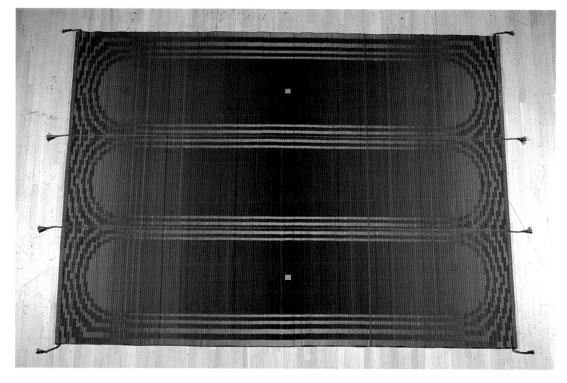

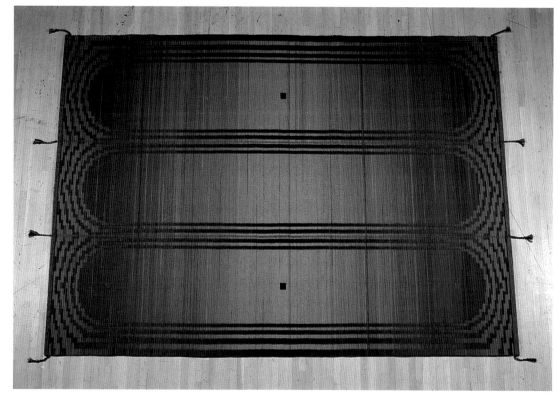

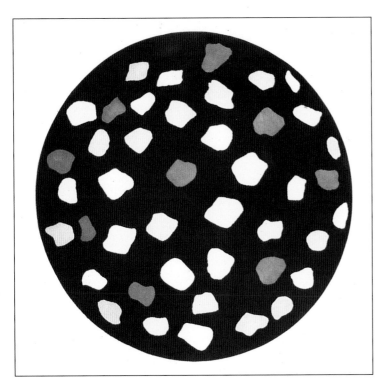

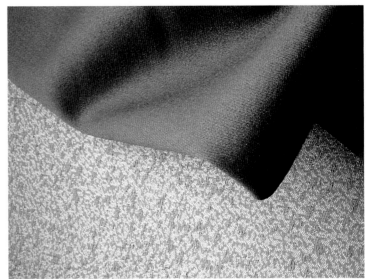

PRODUCT
Provénce Collection (Calypso and Montage)
DESIGNER
Laura Deubler Mercurio
MANUFACTURER
Adam James Textiles
CLIENT
Adam James Textiles

PRODUCT
Patterned Area Rug
DESIGNER
Michael McDonough
FIRM
Michael McDonough Studio
MANUFACTURER
Galaxy Carpet Mills
DESCRIPTION
Coloration and patterning are inspired by Zuni and other native American ceremonial objects

PRODUCT
Provénce Collection (Nuance)
DESIGNER
Laura Deubler Mercurio
MANUFACTURER
Adam James Textiles
CLIENT
Adam James Textiles

PRODUCT
St. Ives Carpet
DESIGNER
Flori Hendron
MANUFACTURER
Patcraft Carpet Mills
PHOTOGRAPHY
Jim Sullivan

PRODUCT
Bolidist Rug
DESIGNER
Massimo Iosa-Ghini
MANUFACTURER
Palazzetti

PRODUCT
"Cabernet Prints" wallcovering
MANUFACTURER
J.M. Lynne Co., Inc.
CLIENT
J.M. Lynne Co., Inc.

PRODUCT
New Ways
DESIGNER
Richard Giglio
FIRM
Donghia
DESCRIPTION
Four patterns on cotton
pebblecloth are based on the
paintings, drawings and collages
of artist Richard Giglio
CLIENT
Donghia Textiles
PHOTOGRAPHY
Donghia

45

PRODUCT
Handprinted fabrics
DESIGNER
Joel Sokolov

PRODUCT
Provénce Collection (Peacock)
DESIGNER
Laura Deubler Mercurio
MANUFACTURER
Adam James Textiles
CLIENT
Adam James Textiles

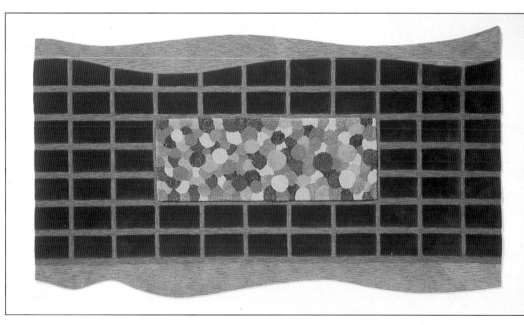

PRODUCT
New York, New York Rug
DESIGNER
Frederic Schwartz
FIRM
Anderson/Schwartz Architects
MANUFACTURER
V'Soske
DESCRIPTION
Idealized look of Manhattan as a
floor map

PRODUCT
The Earth
DESIGNER
Doomey
MANUFACTURER
Doomey
DESCRIPTION
Spread in leather, six skins are
sewed together. Each represents
a continent and is painted with
symbols of the continent.
CLIENT
Le Court-Bredat
PHOTOGRAPHY
Doomey

PRODUCT
Provénce Collection (Moonlight)
DESIGNER
Laura Deubler Mercurio
MANUFACTURER
Adam James Textiles
CLIENT
Adam James Textiles

PRODUCT
Textiles
DESIGNER
Nina Sobell
DESCRIPTION
Handprinted dyes on silk

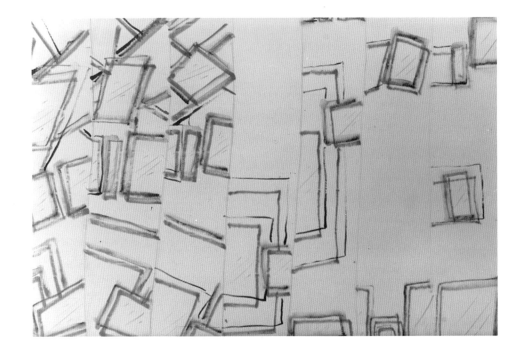

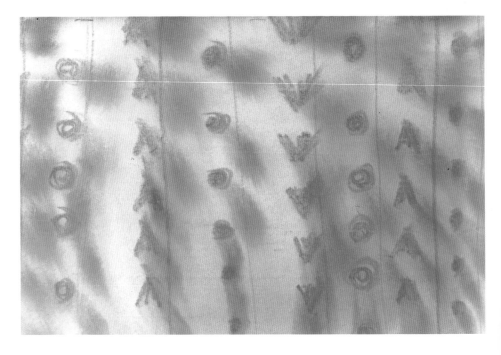

PRODUCT
"Mother and Child"
DESIGNER
Audrey Goldstein
DESCRIPTION
Hand-painted textile panel
PHOTOGRAPHY
Mark Diamond

PRODUCT
"Sketchbook Series #3"
DESIGNER
Audrey Goldstein
DESCRIPTION
Hand-painted textile panel
PHOTOGRAPHY
Mark Diamond

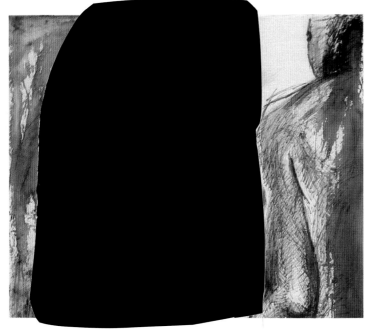

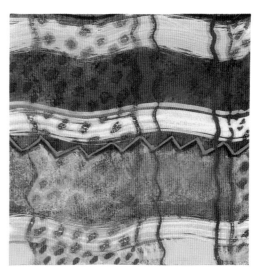

PRODUCT
"Moving Around"
DESIGNER
Audrey Goldstein
DESCRIPTION
Hand-painted textile panel
PHOTOGRAPHY
Mark Diamond

PRODUCT
"Zig Zag Plaid" Textile
DESIGNER
Audrey Goldstein
PHOTOGRAPHY
Mark Diamond

3 Furniture

INDIVIDUAL IDENTITY HAS BEEN lost to the status symbol. The Pepsi generation and the BMW crowd are two such examples. Personal identity turns out to be group identification: class, ethnic, design-consciousness, etc. So, if being hip is reflected in one's clothes or car, then please be truly hip – avoid oppression based on what products other people use. Beware: the Marlboro Man supports Jesse Helms, so understand the repercussions of your brand. "New" may show off purchasing power, but "old and well maintained" shows off much more. Go look at the garbage dump before you opt for disposable. We are also the products we throw out.

As a designer in home furnishings, I use the handmade/limited edition approach as an antidote to the mass produced world we live in. Simple primitive shapes and patterns, put together in a relaxed manner, can be a counterbalance to the stamped out, clean look of appliances and electronics. In home design, references to history, culture and even ritual, enrich the domestic landscape. "Bienvenido" table (meaning "welcome") marks the return from a ⟨...⟩ with new

materials broadens the scope of tools that can be used in product design.

In the future, design will react more strongly against the mass-produced world we live in. People will need to feel even more human, and less as a consumer. Jetsons merge with Flintstones.

DON RUDDY
Designer

Don Ruddy's clients, who come from fields as diverse as retail, graphics, restaurants, interiors and product design, include Horchow, Zona, Anne Klein Knitwear, International Design Group, London Records, Martex, Arcadia restaurant and Café Luxembourg.

Based in New York, Mr. Ruddy has displayed his designs at the Parson School of Design, Max Protetch Gallery, International Contemporary Furniture Fair and the Metropolitan Museum of Art. His work has also been published in the New York Times, Interiors, Industrial Design, Metropolis and Metropolitan Home.

Mr. Ruddy attended the School of Environmental Studies at University College in London, and has a degree in Architecture from the University of Virginia.

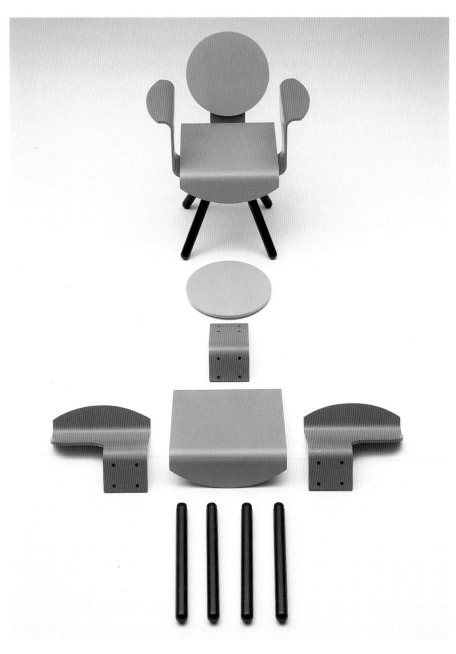

PRODUCT
"August Chair"
DESIGNER
Shigeru Uchida
FIRM
Studio 80
MANUFACTURER
Chairs
PHOTOGRAPHY
Nacása & Partners Inc.

PRODUCT
"August" Stool
DESIGNER
Shigeru Uchida
FIRM
Studio 80
MANUFACTURER
Chairs
PHOTOGRAPHY
Nacása & Partners Inc.

PRODUCT
Arne and Arnochair
(Beast Collection)
DESIGNER
David Shaw Nicholls
FIRM
David Shaw Nicholls Corp.
DISTRIBUTOR
Modern Age

PRODUCT
"Glasnost" stool
DESIGNER
Maurizio Peregalli
CLIENT
Noto
PHOTOGRAPHY
Bitetto-Chimenti

PRODUCT
"Front Row Center" Chair 1989
DESIGNER
James Evanson
MANUFACTURER
James Evanson and Art et
Industrie
DESCRIPTION
Constructed of wood, copper leaf
and cassette player

53

PRODUCT
Damas
DESIGNER
Jeannot Cerutti
FIRM
Ceccotti Collezione
MANUFACTURER
Ceccotti Collezione
DESCRIPTION
Armchair of solid padauk wood
with a wax finish
CLIENT
Ceccotti Collezione
PHOTOGRAPHY
Mario Ciampi
DISTRIBUTOR
Frederic Williams

PRODUCT
Easy Chair
DESIGNER
Will Stone
FIRM
Lewis Dolin, Inc.
MANUFACTURER
Will Stone
PHOTOGRAPHY
Jason Jones

PRODUCT
School Buddies
DESIGNER
Tom Freedman
FIRM
Cutting Edge
MANUFACTURER
Cutting Edge
PHOTOGRAPHY
Tom Freedman

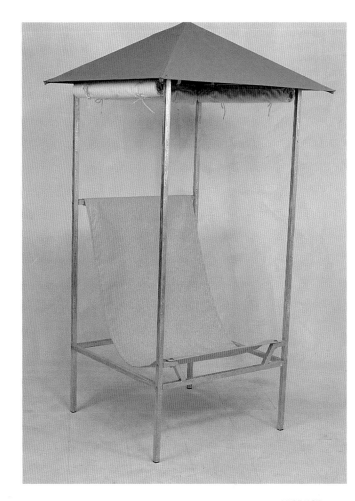

PRODUCT
Cabana
DESIGNER
Allen Miesner
FIRM
Miesner Design

PRODUCT
D.R.D.P. (Double rêve du printemps)
DESIGNER
Roberto Lazzeroni
FIRM
Ceccotti Collezione
MANUFACTURER
Ceccotti Collezione
DESCRIPTION
Two-seater settee — frame in solid cherry wood with a wax finish; upholstered seat with polyurethane foam covered with velvet; small tray has a silver top
CLIENT
Ceccotti Collezione
PHOTOGRAPHY
Mario Ciampi
DISTRIBUTOR
Frederic Williams

PRODUCT
Hypnos
DESIGNER
Roberto Lazzeroni
FIRM
Ceccotti Collezione
MANUFACTURER
Ceccotti Collezione
DESCRIPTION
Armchair — frame in solid cherry wood with a wax finish; upholstered seat with polyurethane foam covered in velvet.
CLIENT
Ceccotti Collezione
PHOTOGRAPHY
Mario Ciampi
DISTRIBUTOR
Frederic Williams

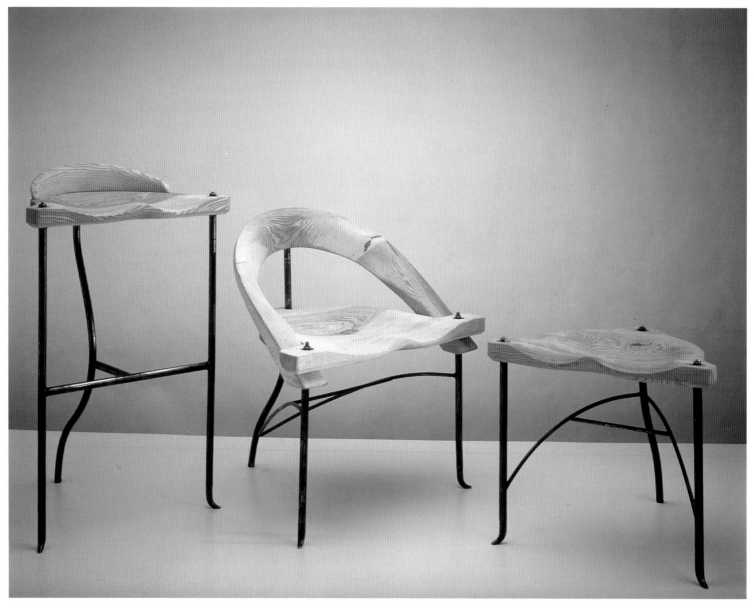

PRODUCT
Noah Collection
DESIGNER
Nigel Coates
MANUFACTURER
Palazzetti

PRODUCT
ZOS Lounge Chair
DESIGNER
John Beckmann
DISTRIBUTOR
Axis Mundi Inc.

PRODUCT
"Twiddler" Chair 1989
DESIGNER
Terence Main
MANUFACTURER
Terence Main and Art et Industrie

PRODUCT
Ciripá
DESIGNER
Maurizio Peregalli
CLIENT
Noto
PHOTOGRAPHY
Bitetto-Chimenti

PRODUCT
Scarborough Chairs Numbers 22
and 8 with Telephone Table on
Crutches
DESIGNER
Dane Scarborough
MANUFACTURER
Dane Scarborough
DESCRIPTION
Of solid ash, Italian bending
poplar and steel
PHOTOGRAPHY
Landis McIntire
DISTRIBUTOR
Tim Wells Furniture

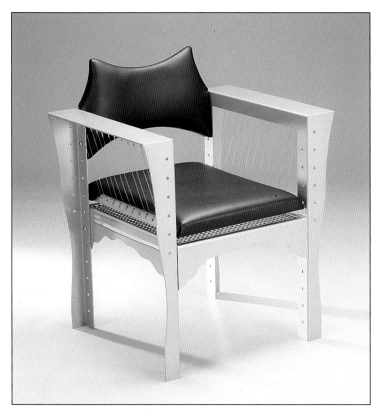

PRODUCT
Cosmic Ray Chair
DESIGNER
Shozo Toyohisa
FIRM
Eastern Accent International Inc.
DESCRIPTION
Made of iron, leather and stainless wire
DISTRIBUTOR
Eastern Accent International Inc.

PRODUCT
Formal Chair
DESIGNER
Paul Bradley
FIRM
Matrix Product Design
DESCRIPTION
Constructed of mahogany
CLIENT
Object Design
PHOTOGRAPHY
Rick English

PRODUCT
Finestra
DESIGNER
Michael Graves
FIRM
Atelier International, Ltd.
MANUFACTURER
Atelier International, Ltd.
DESCRIPTION
Wood frame pull-up chair
PHOTOGRAPHY
Studio photographers in Italy

PRODUCT
Oculus Armchair
DESIGNER
Michael Graves
FIRM
Atelier International, Ltd.
MANUFACTURER
Atelier International, Ltd.
DESCRIPTION
Wood frame pull-up chair
PHOTOGRAPHY
Studio photographers in Italy

PRODUCT
Informal Chair
DESIGNER
Paul Bradley
FIRM
Matrix Product Design
DESCRIPTION
Maple, urethane foam, lacquer furniture
CLIENT
Object Design
PHOTOGRAPHY
Rick English

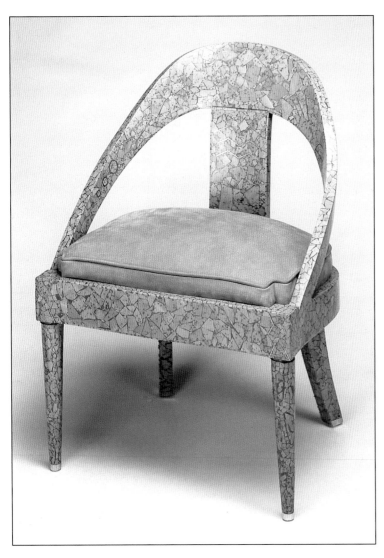

PRODUCT
Loisaida Chair
DESIGNER
Iris DeMauro
FIRM
GEO International
CLIENT
GEO International
PHOTOGRAPHY
Victor Schrager

PRODUCT
Maquette for Folding Writing
Desk
DESIGNER
David Deutsch
FIRM
A/D
DESCRIPTION
Standing, folding writing desk to
be fabricated of aluminum
tubing, with aluminum writing
surface and footrest, with vinyl
webbing
PHOTOGRAPHY
Ken Schles
DISTRIBUTOR
A/D

PRODUCT
Musical Chairs 1989
DESIGNER
Lloyd Schwan
FIRM
Godley-Schwan
DESCRIPTION
Side chairs

PRODUCT
Vanadon Chair
DESIGNER
James VanEtten
FIRM
Delbanco Arts

PRODUCT
Celestial Chair 1989
DESIGNER
Michele Oka Doner
MANUFACTURER
Michele Oka Doner and Art et
Industrie
DESCRIPTION
Constructed of bronze

PRODUCT
Im-pax chair
DESIGNER
Maurizio Peregalli
CLIENT
Noto
PHOTOGRAPHY
Bitetto-Chimenti

61

PRODUCT
Midra
DESIGNER
Maurizio Peregalli
CLIENT
Noto
PHOTOGRAPHY
Bitetto-Chimenti

PRODUCT
Baisity Seating Systems
DESIGNER
Antonio Citterio
MANUFACTURER
B & B Italia
PHOTOGRAPHY
Aldo Ballo

PRODUCT
Chair of the Future
DESIGNER
Shozo Toyohisa
FIRM
Eastern Accent International Inc.
DESCRIPTION
Mixed media chair: marble,
leather, brass and steel
DISTRIBUTOR
Eastern Accent International Inc.

PRODUCT
"January" Sofa
DESIGNER
Shigeru Uchida
FIRM
Studio 80
MANUFACTURER
Chairs
PHOTOGRAPHY
Nacása & Partners Inc.

PRODUCT
St. James™ Club Chair
FIRM
Donghia
DESCRIPTION
High throne-like back and elongated arms for high style and comfort
PHOTOGRAPHY
Donghia

PRODUCT
Baisity Seating Systems
DESIGNER
Antonio Citterio
MANUFACTURER
B & B Italia
PHOTOGRAPHY
Aldo Ballo

PRODUCT
"Spanner" Chair
DESIGNER
David Mocarski
FIRM
Taction Design
DESCRIPTION
Armchair and footstool

PRODUCT
Hollington Chair
DESIGNER
Geoff Hollington
FIRM
Hollington Associates
MANUFACTURER
Herman Miller Inc. U.S.A.
DESCRIPTION
Lounge chair and ottoman
CLIENT
Herman Miller Inc. U.S.A.

PRODUCT
Royalton Bar Stool
DESIGNER
Philippe Starck
MANUFACTURER
XO
DISTRIBUTOR
Modern Age

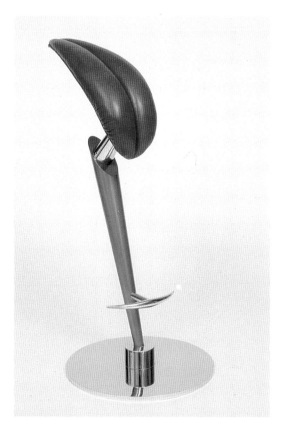

PRODUCT
Orissor stool
DESIGNER
John Beckmann
DISTRIBUTOR
Axis Mundi Inc.

PRODUCT
"Modern G" barstool
DESIGNER
David Mocarski
FIRM
Taction Design
PHOTOGRAPHY
Charles Imstepf

PRODUCT
Y.M.D. IMONO stool
DESIGNER
Takenobu Igarashi
FIRM
Igarashi Studio
MANUFACTURER
Yamasho Casting
DESCRIPTION
Created by using traditional cast
iron methods used in Yamagata
Prefecture for over 900 years
PHOTOGRAPHY
Masaru Mera
DISTRIBUTOR
Yamada Shomei Lighting Co.,
Y.M.D. Division

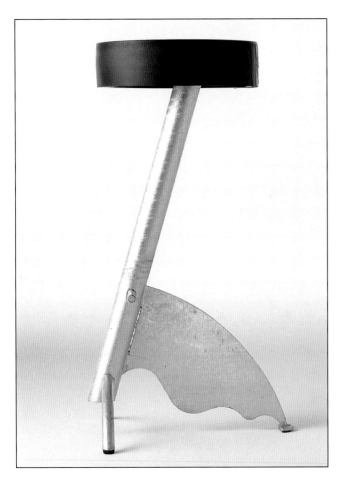

PRODUCT
"Platform" stool
DESIGNER
Maurizio Peregalli
CLIENT
Noto
PHOTOGRAPHY
Bitetto-Chimenti

PRODUCT
Chair of Forgiveness
DESIGNER
Richard Snyder
FIRM
Richard Snyder Design
DESCRIPTION
Four-posted chair with silk cushion
and multicolored tassels: "Sit
once and you will forgive all!"
PHOTOGRAPHY
Bill White

PRODUCT
Ribbon Chaise
DESIGNER
Gary Stephen
FIRM
A/D
DESCRIPTION
Of mahogany, brass caning and
black leather
PHOTOGRAPHY
Ken Schles
DISTRIBUTOR
A/D

66

PRODUCT
"Sleepers" Chaise Longue 1989
DESIGNER
Alex Locadia
MANUFACTURER
Alex Locadia and Art et Industrie

PRODUCT
Laurel Leaf Sofa
DESIGNER
David Davies
MANUFACTURER
David Davies Association for
Luten Clarey Stern Inc.

PRODUCT
Baisity Seating Systems
DESIGNER
Antonio Citterio
MANUFACTURER
B & B Italia
PHOTOGRAPHY
Aldo Ballo

PRODUCT
Baisity Seating Systems
DESIGNER
Antonio Citterio
MANUFACTURER
B & B Italia
PHOTOGRAPHY
Aldo Ballo

PRODUCT
Sofa
DESIGNER
Jasper Morrison
MANUFACTURER
Palazzetti

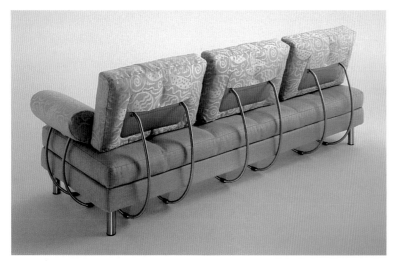

PRODUCT
Europa Sofa
DESIGNERS
Gualtierotti and Mazzoni delle
Stelle
MANUFACTURER
Zanotta s.p.a.
DISTRIBUTOR
Modern Age

PRODUCT
Sity (seating system)
DESIGNER
Antonio Cittrerio
MANUFACTURER
B & B Italia
PHOTOGRAPHY
G. Pierre Maurer

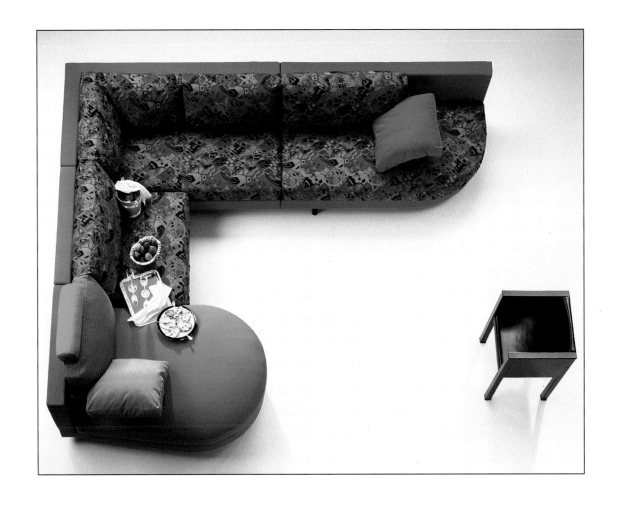

69

PRODUCT
Table
DESIGNER
Joseph Martell
FIRM
A/D
DESCRIPTION
Constructed with Rosso levanto
marbles and glass
PHOTOGRAPHY
Ken Schles
DISTRIBUTOR
A/D

PRODUCT
Galaxy Table
DESIGNER
Iris DeMauro
FIRM
GEO International
CLIENT
GEO International
PHOTOGRAPHY
Victor Schrager

PRODUCT
Connecting VI Dining Table
DESIGNER
Stanley Weksler Casselman
FIRM
Casselglass International
DESCRIPTION
Constructed with green Monolith glass™

PRODUCT
Cities Table Collection
DESIGNER
Al Glass
FIRM
Becker Designed, Inc.
MANUFACTURER
Becker Designed, Inc.
DESCRIPTION
Cordovan-stained cherry and steel
PHOTOGRAPHY
Len Rizzi

PRODUCT
Delbanco Table
DESIGNER
Kurt Delbanco
FIRM
Delbanco Arts
PHOTOGRAPHY
John Schwartz

PRODUCT
Lin Table
DESIGNER
Demir Hamami
FIRM
Becker Designed, Inc.
MANUFACTURER
Becker Designed, Inc.
DESCRIPTION
Steel frame with perforated and
anodized aluminum screen and
glass top.
PHOTOGRAPHY
Len Rizzi

PRODUCT
"L'armadio per Aldo" Cabinet
DESIGNER
Shigeru Uchida
FIRM
Studio 80
MANUFACTURER
Chairs
DESCRIPTION
Made of birch plywood with
pebble finish; legs of cherry wood
with adjustable worm screws
PHOTOGRAPHY
Nacása & Partners Inc.

PRODUCT
Il Tavolo Per Aldo
DESIGNER
Shigeru Uchida
FIRM
Studio 80
MANUFACTURER
Chairs
DESCRIPTION
Made from birch plywood with
pebble finish, legs of cherry wood
with adjustable worm screws
PHOTOGRAPHY
Nacása & Partners Inc.

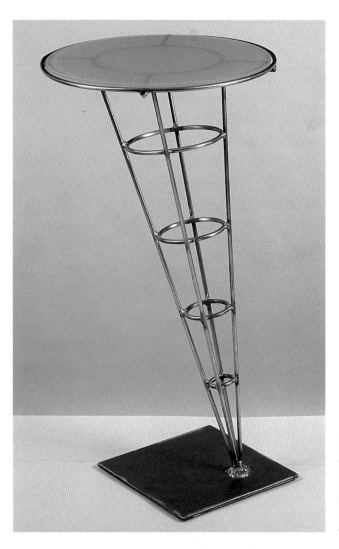

PRODUCT
Woven Bench
DESIGNER
David Hess
FIRM
Lewis Dolin, Inc.
MANUFACTURER
David Hess
DESCRIPTION
Composed of galvanized steel weave and anchored by rebar
PHOTOGRAPHY
David Hess

PRODUCT
"Cobalt Series—Delineations in Steel" Unicone End Table
DESIGNER
Peter Diepenbrock
MANUFACTURER
Peter Diepenbrock
PHOTOGRAPHY
James Beards

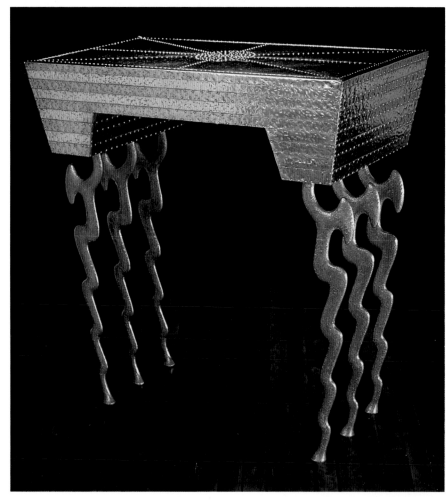

PRODUCT
Console (from Oteko Collection)
DESIGNER
Peter Diepenbrock
MANUFACTURER
Peter Deipenbrock
PHOTOGRAPHY
James Beards

PRODUCT
Flipper Table
DESIGNER
Matthew Hilton
MANUFACTURER
Palazzetti

PRODUCT
Tovaglia Table 1989
DESIGNER
Davide Mercatali
DESCRIPTION
Constructed of natural steel and
a yellow zinc metallized
CLIENT
METALS
PHOTOGRAPHY
Sergio Merli

PRODUCT
Tom Tom Tables
DESIGNER
John Eric Byers
MANUFACTURER
John Eric Byers
DESCRIPTION
Base of ebonized cherry, tops of various wood veneers
PHOTOGRAPHY
Tom Brummett

PRODUCT
Hemisphere Table
MANUFACTURER
Brueton Industries

PRODUCT
"Bienvenido" Table
DESIGNER
Don Ruddy
MANUFACTURER
Don Ruddy
DESCRIPTION
Made of pigmented concrete
with inlaid glass and tiles
CLIENT
Edgardo Heyra

PRODUCT
Tadao
DESIGNER
Roberto Lazzeroni
FIRM
Ceccotti Collezione
MANUFACTURER
Ceccotti Collezione
DESCRIPTION
Ovoid-shaped sideboard — solid
cherry wood or walnut with a
wax finish, two doors and two
drawers; internal part of drawers
and internal shelves in maple and
padauk respectively.
CLIENT
Ceccotti Collezione
PHOTOGRAPHY
Mario Ciampi
DISTRIBUTOR
Frederic Williams

PRODUCT
Table
DESIGNER
David Perry
DESCRIPTION
Constructed of wood and copper
PHOTOGRAPHY
Susan Einstein

PRODUCT
"Twilight" Table 1989
DESIGNER
Howard Meister
MANUFACTURER
Howard Meister and Art Et
Industrie
DESCRIPTION
Hand-wrought and made of
painted steel and glass

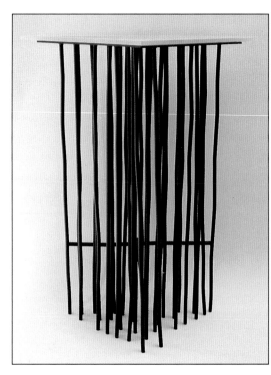

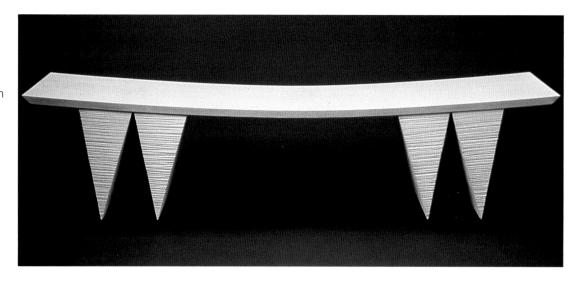

PRODUCT
Garden Bench
DESIGNER
John Eric Byers
MANUFACTURER
John Eric Byers
DESCRIPTION
Constructed of painted ash
PHOTOGRAPHY
Tom Brummett

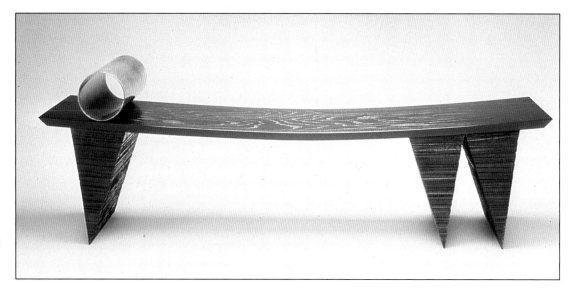

PRODUCT
Urban Chaise
DESIGNER
John Eric Byers
MANUFACTURER
John Eric Byers
DESCRIPTION
Constructed of ebonized ash and
patinated copper
PHOTOGRAPHY
Tom Brummett

PRODUCT
Phoenician Magician's Chest
DESIGNER
Richard Snyder
FIRM
Richard Snyder Design
DESCRIPTION
Sideboard with gull wing doors
on top and whitewashed interior
PHOTOGRAPHY
Bill White

PRODUCT
"Diamond Ladder" Dining Table
DESIGNER
Laura Johnson
MANUFACTURER
Laura Johnson and Art et
Industrie
DESCRIPTION
Constructed of slate, carved
wood, lacquer and inlaid copper

PRODUCT
Table and 6 Chairs
DESIGNER
Joel Sokolov

PRODUCT
Copper Chaise
DESIGNER
David Hess
FIRM
Lewis Dolin, Inc.
MANUFACTURER
David Hess
DESCRIPTION
Composed of copper weave and
rebar
PHOTOGRAPHY
Lewis Dolin

80

PRODUCT
Circle Table Collection
DESIGNER
Bill Becker
FIRM
Becker Designed, Inc.
MANUFACTURER
Becker Designed, Inc.
DESCRIPTION
Cocktail table and console table
with sandblasted glass supported
by steel columns
PHOTOGRAPHY
Len Rizzi

PRODUCT
Ariannetable (Beast Collection)
DESCRIPTION
David Shaw Nicholls
FIRM
David Shaw Nicholls Corp.
DISTRIBUTOR
Modern Age

PRODUCT
TV/Stereo Cabinet
DESIGNER
John Eric Byers
MANUFACTURER
John Eric Byers
DESCRIPTION
Cabinet on arch
PHOTOGRAPHY
Tom Brummett

PRODUCT
Table Figure
DESIGNER
John Eric Byers
MANUFACTURER
John Eric Byers
DESCRIPTION
Ebonized ash table and mirror
frame
PHOTOGRAPHY
Tom Brummett

PRODUCT
Table Figure
DESIGNER
John Eric Byers
MANUFACTURER
John Eric Byers
DESCRIPTION
Ebonized ash table and mirror
frame
PHOTOGRAPHY
Tom Brummett

PRODUCT
Arabella
DESIGNER
Fabrizia Scassellati Sforzolini
FIRM
Ceccotti Collezione
MANUFACTURER
Ceccotti Collezione
DESCRIPTION
Small writing desk — solid
mahogany with a wax finish,
central drawer and two side
swivel trays
CLIENT
Ceccotti Collezione
PHOTOGRAPHY
Mario Ciampi
DISTRIBUTOR
Frederic Williams

PRODUCT
Sideboard
DESIGNER
Tim Wells
FIRM
Fred Baier and Tim Wells
Partnership
MANUFACTURER
Tim Wells, The Pressure Group
DESCRIPTION
Constructed of English sycamore
veneer and stained maple
PHOTOGRAPHY
David Mohney
DISTRIBUTOR
Tim Wells Furniture

PRODUCT
"Paralelas" Management Table
(full view)
DESIGNER
Jaime Tresserra
FIRM
J. Tresserra Design S.L.

PRODUCT
Buck Paralelas auxiliary table
DESIGNER
Jaime Tresserra
FIRM
J. Tresserra Design S.L.

PRODUCT
BUCKS Paralelas/Tensor Auxiliary
Tables
DESIGNER
Jamie Tresserra
FIRM
J. Tresserra Designs S.L.

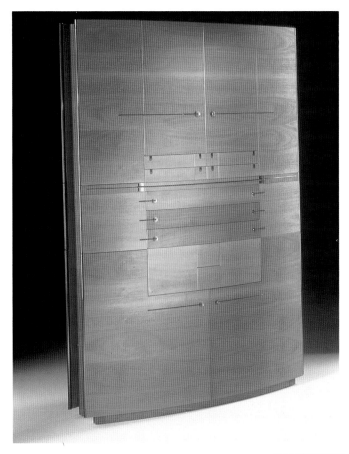

PRODUCT
"Samuro" (full view, closed)
DESIGNER
Jaime Tresserra
FIRM
J. Tresserra Design S.L.
AWARD
SIDI Selection (Best Furniture)

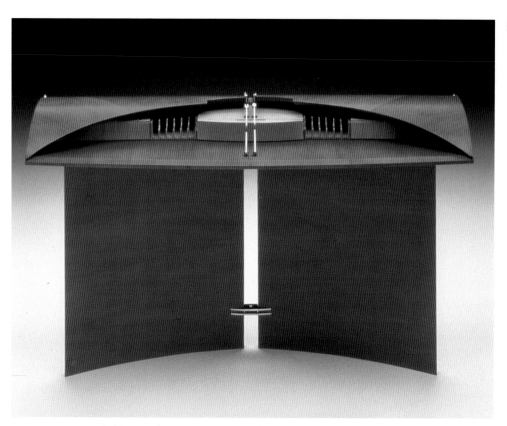

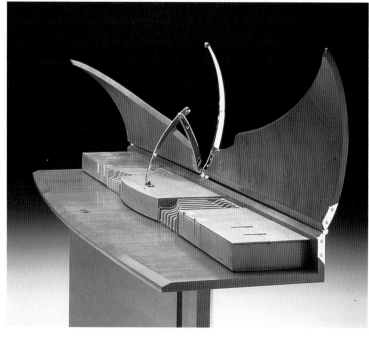

PRODUCT
Green Desk
DESIGNER
Beth Yoe
FIRM
Cutting Edge
MANUFACTURER
Cutting Edge
PHOTOGRAPHY
Tom Freedman

PRODUCT
Domus Container-Based
Furnishings
DESIGNER
Antonio Citterio
MANUFACTURER
B & B Italia
DESCRIPTION
Bookcases, showcases, wall
fittings
PHOTOGRAPHY
Aldo Ballo

PRODUCT
"Clever Moments/Quicker Than
The Eye" Desk (full view)

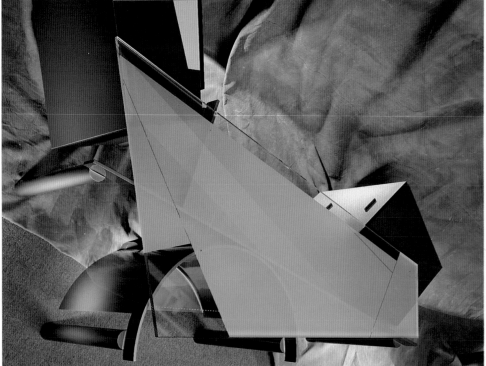

PRODUCT
"Clever Moments/Quicker Than
The Eye" Desk (aerial view)
DESIGNER
David Mocarski
FIRM
Taction Design
CLIENT
Karl Bornstein
PHOTOGRAPHY
Charles Imstepf

PRODUCT
"Blind-Sided/Watching Over You"
Desk
DESIGNER
David Mocarski
FIRM
Taction Design
CLIENT
Peter Weil
PHOTOGRAPHY
Charles Imstepf

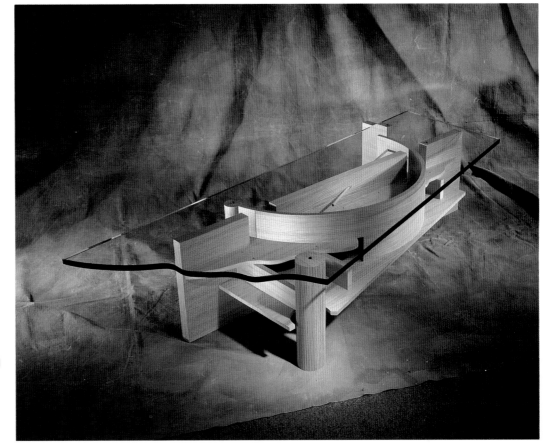

PRODUCT
"Once Upon A Time/Thunder and
Lightning" Coffee Table
DESIGNER
David Mocarski
FIRM
Taction Design
PHOTOGRAPHY
Charles Imstepf

PRODUCT
Domus Container-Based
Furnishings
DESIGNER
Antonio Citterio
MANUFACTURER
B & B Italia
DESCRIPTION
Bookcases, showcases, wall
fittings
PHOTOGRAPHY
Aldo Ballo

PRODUCT
Console
DESIGNER
Alan S. Kushner
DESCRIPTION
Constructed of colored mahogany, zebra wood, ebony and marble
PHOTOGRAPHY
Alice Sebrell

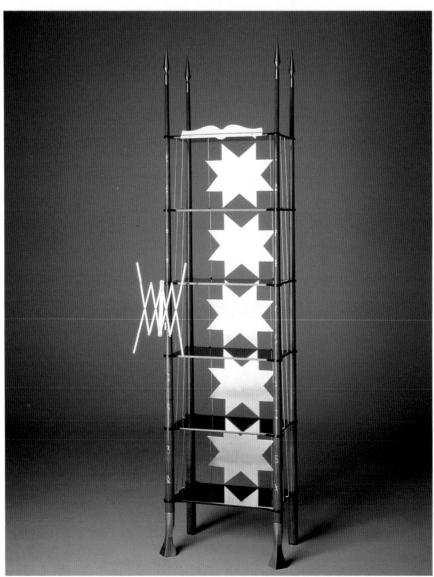

PRODUCT
Libabel Bookcase
MANUFACTURER
Sawaya & Moroni
DISTRIBUTOR
Modern Age

PRODUCT
Citi-cabinet 2 (open)

PRODUCT
Citi-cabinet 2
DESIGNER
Joel Sokolov

PRODUCT
Dowry Chest
DESIGNER
John Eric Byers
MANUFACTURER
John Eric Byers
DESCRIPTION
Constructed of ebonized oak and
ash with patinated copper rivets
PHOTOGRAPHY
Tom Brummett

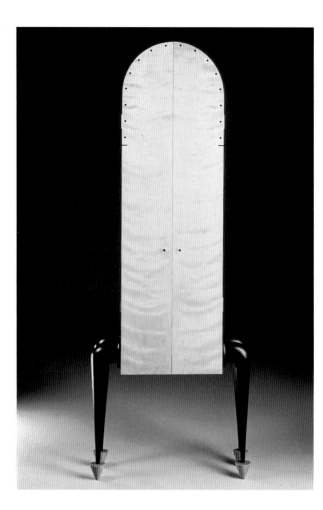

PRODUCT
High-Boy Cabinet
DESIGNER
Dale Broholm
MANUFACTURER
Dale Broholm
PHOTOGRAPHY
Powell Photography

PRODUCT
Highboy
DESIGNER
John Eric Byers
MANUFACTURER
John Eric Byers
DESCRIPTION
Constructed of ebonized oak,
mahogany and ash with
patinated copper rivets
PHOTOGRAPHY
Tom Brummett

PRODUCT
Lowboy
DESIGNER
John Eric Byers
MANUFACTURER
John Eric Byers
DESCRIPTION
Constructed of ebonized oak,
mahogany and curly maple
PHOTOGRAPHY
Tom Brummett

PRODUCT
Cabinet of Four Wishes
DESIGNER
Richard Snyder
FIRM
Richard Snyder Design
DESCRIPTION
A roughly-textured ancient-looking chest of drawers, in lacquered mahogany with brass
PHOTOGRAPHY
Bill White

PRODUCT
3 Units
DESIGNER
Joel Sokolov

PRODUCT
Onda Quadra
DESIGNER
Mario Bellini
FIRM
Atelier International, Ltd.
MANUFACTURER
Atelier International, Ltd.
DESCRIPTION
Stacked storage units
PHOTOGRAPHY
Studio photographers in Italy

PRODUCT
Onda Quadra
DESIGNER
Mario Bellini
FIRM
Atelier International, Ltd.
MANUFACTURER
Atelier International, Ltd.
DESCRIPTION
Stacked storage units
PHOTOGRAPHY
Studio photographers in Italy

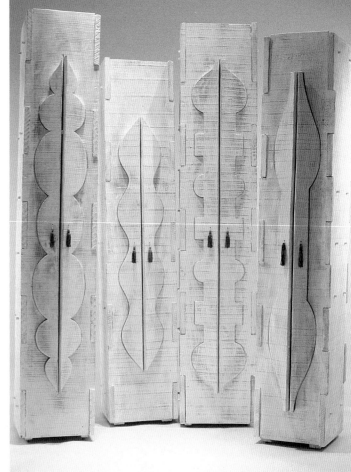

PRODUCT
Fertility Cabinets
DESIGNER
Richard Snyder
FIRM
Richard Snyder Design
DESCRIPTION
Vertical wood cabinets with
irregularly shaped doors and
antique tassels
PHOTOGRAPHY
Joe Coscia

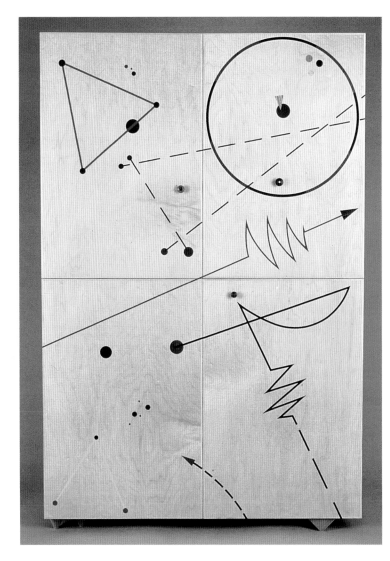

PRODUCT
"Wedge" Desk
DESIGNER
Allen Miesner
FIRM
Miesner Design

PRODUCT
"Things You See From The Sky #3" Media Cabinet
DESIGNER
Dale Broholm
MANUFACTURER
Dale Broholm
PHOTOGRAPHY
DuBusc Photography

PRODUCT
Love-Slave Box
DESIGNER
Richard Snyder
FIRM
Richard Snyder Design
DESCRIPTION
Sideboard has a secret door to access interior, antique lock on door and brocade cushion inside.
PHOTOGRAPHY
Bill White

PRODUCT
TAU Cabinet
DESIGNER
John Beckmann
MANUFACTURER
Christian Farjon
CLIENT
Christian Farjon
DISTRIBUTOR
Christian Farjon

PRODUCT
Terrain Cabinet 1990
DESIGNER
Lloyd Schwan
FIRM
Godley-Schwan

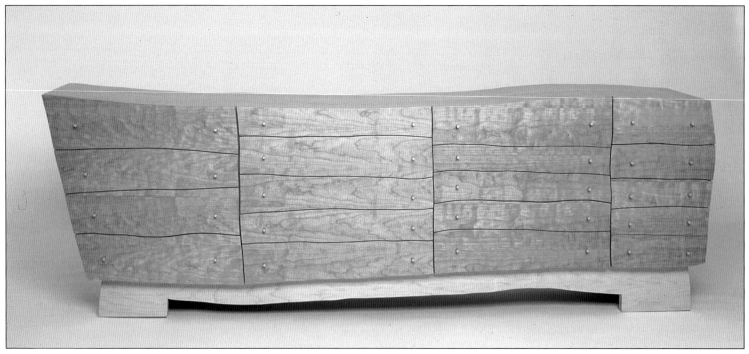

PRODUCT
Citi Cabinet 1
DESIGNER
Joel Sokolov

PRODUCT
Balloon-Shaped Two-Door Display
Cabinet with Lighting
DESIGNER
Alan S. Kushner
PHOTOGRAPHY
Bill Douds

PRODUCT
Astonsideboard (Beast Collection)
DESIGNER
David Shaw Nicholls
FIRM
David Shaw Nicholls Corp.
DISTRIBUTOR
Modern Age

PRODUCT
Mantelpiece
DESIGNER
Beth Forer
MANUFACTURER
Cesar Cabral, Beth Forer
DESCRIPTION
Several patterns made with
neiikomi technique — wirecut
layered colored clay under a clear
glaze.
PHOTOGRAPHY
Beth Forer

PRODUCT
Peaked roof chest
DESIGNER
Beth Forer
MANUFACTURER
Cesar Cabral, Beth Forer
DESCRIPTION
Constructed of handmade tiles
on wood; decorated by Mishima
technique — colored clay inlaid
with contrasting slip under a
clear glaze.
PHOTOGRAPHY
Beth Forer

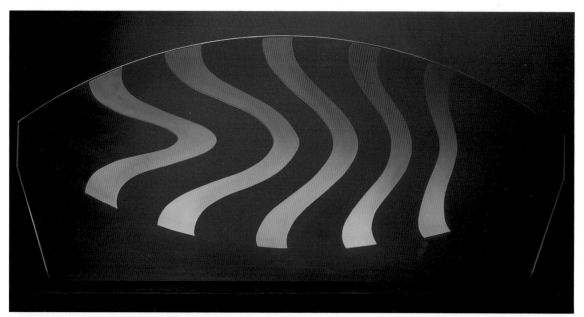

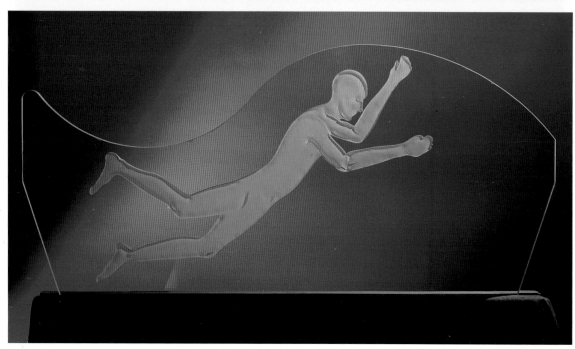

PRODUCT
Headboard
DESIGNER
Baron Bernstein
MANUFACTURER
Baron Bernstein
DESCRIPTION
Queen-sized headboard; shape was grafted on an x/y axis and fed into a computer which controlled a hydrosaber. Frame can appear luminous with neon or other lighting.
PHOTOGRAPHY
William Nettles

PRODUCT
"Big Four Poster" 1989
DESIGNER
Norman Campbell
MANUFACTURER
Norman Campbell and Art et Industrie
DESCRIPTION
Constructed of wood, forged steel and stone

PRODUCT
Regent Bed
DESIGNER
Louis Bromante
FIRM
Sirmos, a division of Bromante
Corp
MANUFACTURER
Sirmos
DESCRIPTION
Inspired by French Regency
period
PHOTOGRAPHY
Sirmos
AWARD
ROSCOE Award

PRODUCT
Steel, Brass, Silicon Bronze Bed
DESIGNER
Philip Miller
CLIENT
Steve Esherman and Cami Taylor
PHOTOGRAPHY
Philip Miller

PRODUCT
Two-panel screen
DESIGNER
Baron Bernstein
MANUFACTURER
Baron Bernstein
DESCRIPTION
Fantasy design in 6' x 2' panels
PHOTOGRAPHY
William Nettles

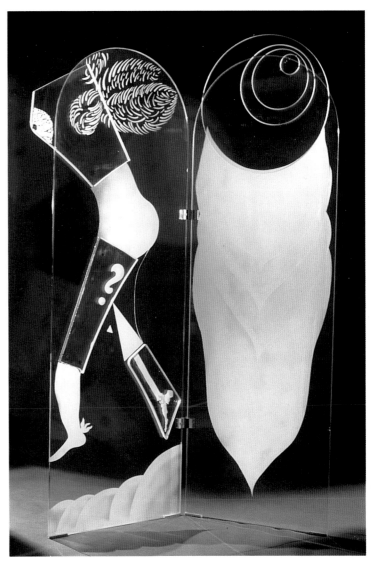

PRODUCT
"Traste" screen
DESIGNER
Jaime Tresserra
FIRM
J. Tresserra Design S.L.

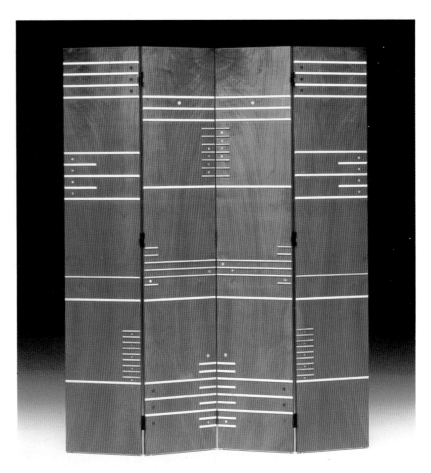

PRODUCT
Silhouette Mirror
DESIGNER
Al Glass
FIRM
Becker Designed, Inc.
MANUFACTURER
Becker Designed, Inc.
DESCRIPTION
Wall mirror with shelf featuring
6mm glass and formed aluminum;
in silver or black enamel.
PHOTOGRAPHY
Len Rizzi

PRODUCT
Mirror (prototype)
DESIGNER
Heiie Damkjaer
PHOTOGRAPHY
Milano Anthrazit

PRODUCT
Leser Mirror Collection
DESIGNER
Max Leser
FIRM
Becker Designed, Inc.
MANUFACTURER
Becker Designed, Inc.
DESCRIPTION
Floor and wall mirrors with 6mm
mirror glass and machined
cylindrical bases
PHOTOGRAPHY
Len Rizzi

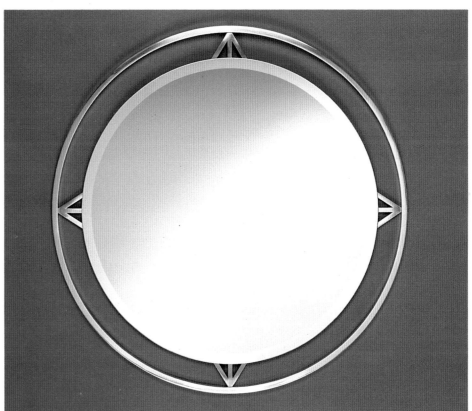

PRODUCT
Etoile Wall Mirror
DESIGNER
David Zelman
FIRM
Prologue 2000
PHOTOGRAPHY
Simon Feldman

PRODUCT
Pezunia
DESIGNER
Pedro Miralles
FIRM
Ceccotti Collezione
MANUFACTURER
Ceccotti Collezione
DESCRIPTION
Hat stands in different types of solid wood (i.e., padauk, cherry, walnut, teak, ash, maple) with a wax finish.
CLIENT
Ceccotti Collezione
PHOTOGRAPHY
Mario Ciampi
DISTRIBUTOR
Frederic Williams

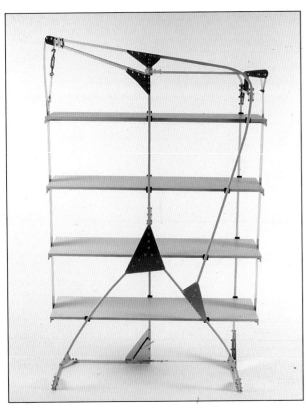

PRODUCT
"du ReBar" Etagere
DESIGNER
Jonathan Teasdale
MANUFACTURER
Jonathan Teasdale and Art et Industrie

PRODUCT
JIGSAW Coat Racks
DESIGNER
John Beckmann; Pouran Esrafily
DESCRIPTION
Fiberglass fishing poles; inter-
locking jigsaw-shaped bases
DISTRIBUTOR
Axis Mundi Inc.

PRODUCT
"Male/Female Coat Rack"
1988/89
DESIGNER
Gloria Kisch
MANUFACTURER
Gloria Kisch and Art et Industrie

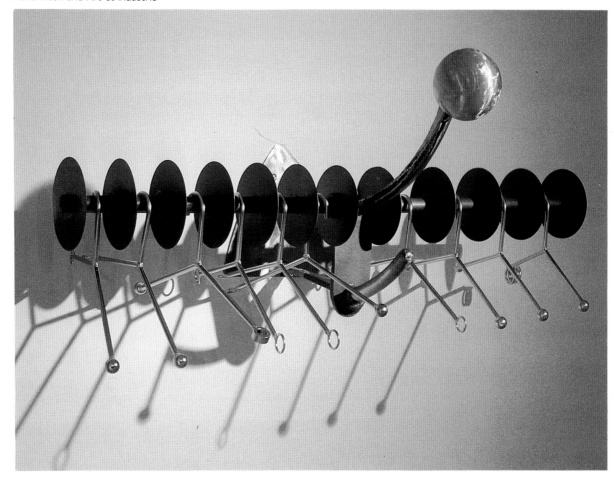

PRODUCT
Ring of Fire
DESIGNER
Richard Snyder
FIRM
Richard Snyder Design
DESCRIPTION
An oversized candelabra of
wrought iron
PHOTOGRAPHY
Bill White

PRODUCT
Baisity Seating Systems
DESIGNER
Antonio Citterio
MANUFACTURER
B & B Italia
DESCRIPTION
Bookcases, showcases, wall
fittings
PHOTOGRAPHY
Aldo Ballo

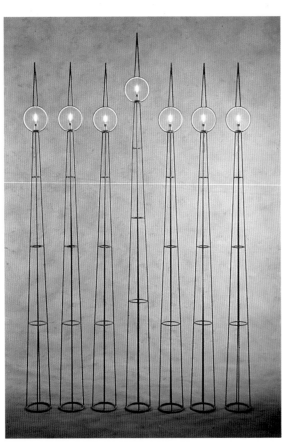

PRODUCT
"Cobalt Series — Delineations in
Steel" Oil Lamps
DESIGNER
Peter Diepenbrock
MANUFACTURER
James Beards
PHOTOGRAPHY
Peter Diepenbrock

PRODUCT
"Dear Morris" Clock
DESIGNER
Shigeru Uchida
FIRM
Studio 80
MANUFACTURER
Chairs
DESCRIPTION
Made of birch plywood with a
corrugated face and pebble finish
PHOTOGRAPHY
Nacása & Partners Inc.

4 Lighting

IF WE ARE THE PRODUCTS WE USE, then to judge by our lamps, most Americans are content with candle light. The vast majority of light fixtures manufactured in the U.S. are "traditional", i.e. based on pre-electric prototypes. Modern is the choice of only a small group of wealthy educated urban consumers, and most of the lighting products they buy are imported from Italy. As a result, it is nearly impossible to get a modern lamp into production here. The public must be enlightened.

I strive to create objects which engage their audience. Surprise, invention and fun are important. The lamps con-

found expectations – in form and materials; often through motion, they amuse and challenge. Anthropomorphic qualities humanize these light machines, and make them personal.

I seek maximum interaction between lamp and user. Choice is designed into the product. In the Hydra lamp, movement allows a user to control the direction and intensity of light, and thus his environment. In the Global Warming lamp, with collaborator Lance Chantry, the modularity allows parts to be interchanged. Thus color, material, and shape may be selected by an owner.

The future will allow fixtures to be individualized even further. Lighting in the coming decade will use high-tech sources for energy efficiency, yet may be decidedly low tech in appearance. Our task is to give new technologies a comprehensible yet expressive form. Our

solutions need be as reductive and emotive as those of the Shakers.

This is an important moment, suspended between the excesses of the '80s and the millennium. We have inherited serious economic and ecological problems. Designers need to take on these challenges. We have an opportunity to reshape the relationship between product, consumer, and environment. We have a responsibility to individuals globally. Let us present a compelling vision of the future.

LEO BLACKMAN
Designer

Leo J. Blackman is an architect and industrial designer working in Manhattan. His furniture and light fixtures are typically assembled from industrial materials (plywood, rubber, plastic, aluminum, Corian and fiberglass) into geometric, often anthropomorphic forms.
    INTERIORS, AMERICAN CRAFT, INDUSTRIAL DESIGN (1985 and 1987 Design Review) and METROPOLITAN HOME MAGAZINE have published Mr. Blackman's work, as have several design-related books. His efforts have been displayed at Gallery 91, the Queens Museum, the Los Angeles Craft Museum and the Brooklyn Museum, which purchased his Blongo Chair for its permanent collection.

PRODUCT
Hemera wall lamp
DESIGNER
Serge Meppiel
CLIENT
Noto
PHOTOGRAPHY
Bitetto-Chimenti

PRODUCT
"Ventosa" floor and table lamp
DESIGNER
Maurizio Peregalli
CLIENT
Noto
PHOTOGRAPHY
Bitetto-Chimenti

PRODUCT
Hemera floor lamp
DESIGNER
Serge Meppiel
CLIENT
Noto
PHOTOGRAPHY
Bitetto-Chimenti

PRODUCT
A3 floor lamp
DESIGNER
Törben Hölmback IDD
FIRM
Hölmback Industrial Design ApS.
MANUFACTURER
BJ Metal AS
DESCRIPTION
The design takes its form from a large sheet of paper
PHOTOGRAPHY
Piotr
DISTRIBUTOR
Lyskilde AS

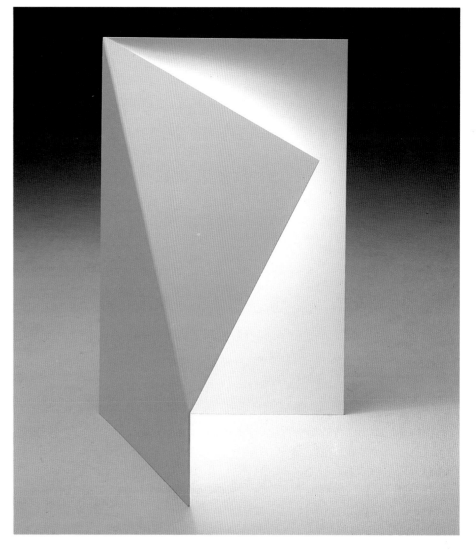

PRODUCT
Sumo
DESIGNER
David Baird
FIRM
Ziggurat
MANUFACTURER
Ziggurat
DESCRIPTION
Table lamp with gold-leafed metal screens and oxidized steel base
CLIENT
Roche-Bobois International, France
PHOTOGRAPHY
Alan Linn

PRODUCT
Global Warming
DESIGNERS
Leo J. Blackman; Lance Chantry
MANUFACTURER
Leo J. Blackman; Lance Chantry
PHOTOGRAPHY
Richard Hackett

PRODUCT
Eddy (wrapped around a tree)
DESIGNER
Noel Zeller
FIRM
Zelco Industries Inc.
MANUFACTURER
Zelco Industries Inc.
DESCRIPTION
Lamp with acrobatic abilities

PRODUCT
Tango
DESIGNER
Stephan Copeland
FIRM
Atelier International, Ltd.
MANUFACTURER
Atelier International, Ltd.
DESCRIPTION
Task lamp adjusts by means of a
uniquely detailed articulated joint.
PHOTOGRAPHY
Studio photographers in Italy

PRODUCT
Zig Zag Lamp
FIRM
Canetti Design Group
MANUFACTURER
Canetti Inc.
DESCRIPTION
Plastic desk/wall lamp
CLIENT
Canetti Inc.
PHOTOGRAPHY
Color Track

PRODUCT
Lashtal Sconce
DESIGNER
John Beckmann; Pouran Esrafily
DESCRIPTION
Silvered-bowl lamp
DISTRIBUTOR
Axis Mundi Inc.

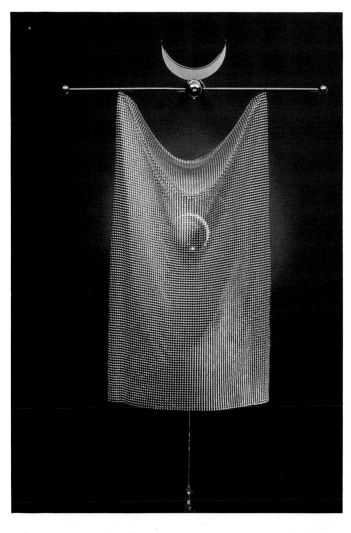

PRODUCT
Hydra Lamp
DESIGNER
Leo J. Blackman
MANUFACTURER
Leo J. Blackman
CLIENT
Leo J. Blackman
PHOTOGRAPHY
Richard Hackett

PRODUCT
"Tria" Torchiere
DESIGNER
Andrzej Duljas
FIRM
Koch + Lowy
MANUFACTURER
Koch + Lowy
PHOTOGRAPHY
Peter Weidlein

PRODUCT
"Ona" Lamp
DESIGNER
Andrzej Duljas
FIRM
Koch + Lowy
MANUFACTURER
Koch + Lowy
PHOTOGRAPHY
Marcus Tullis

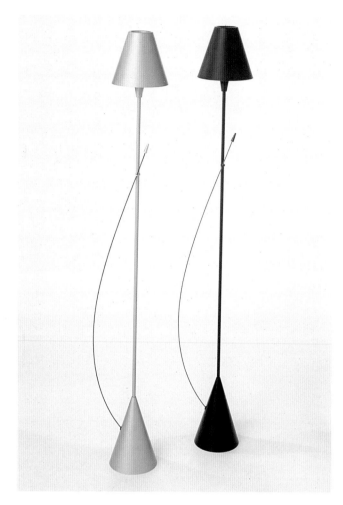

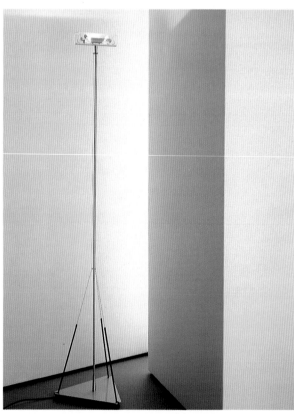

PRODUCT
ACHEO floor
DESIGNER
Gianfranco Frattini
MANUFACTURER
Artemide Inc.
DESCRIPTION
Halogen floor lamp with diffuser
in clear pyrex glass mounted on a
die-cast aluminum support; has
adjustable tension cables
DISTRIBUTOR
Artemide Inc.

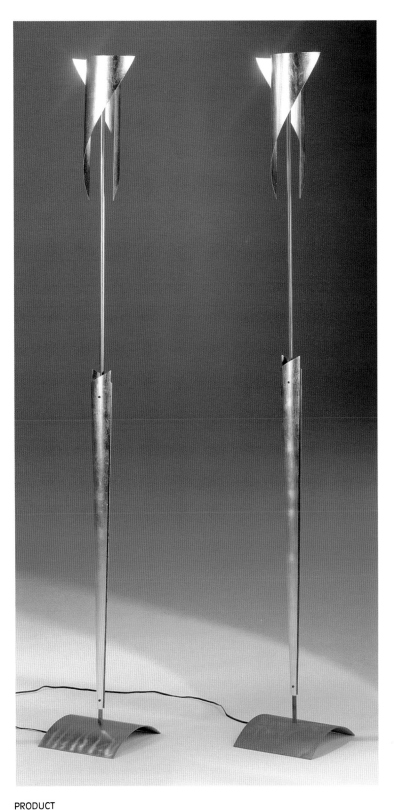

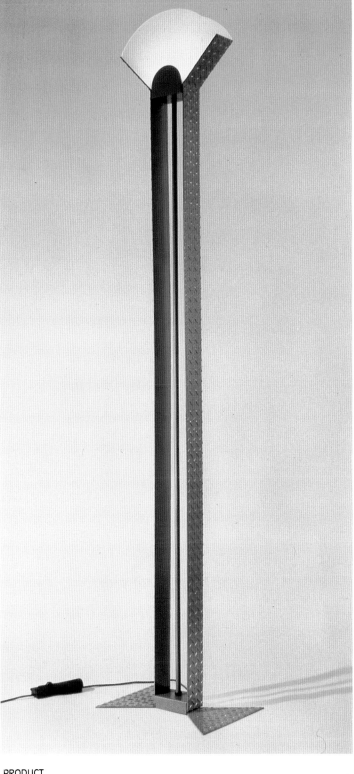

PRODUCT
Veil
DESIGNER
David Baird
FIRM
Ziggurat
MANUFACTURER
Ziggurat
DESCRIPTION
Floor lamp with etched steel
structure and stained wood shelf
CLIENT
Ziggurat
PHOTOGRAPHY
Alan Linn

PRODUCT
Papillon floor lamp
DESIGNER
Roberto Marcatti
FIRM
Lavori in Corso
MANUFACTURER
Ar Far Studio Luce
DESCRIPTION
Lamp in sandblasted buckle
metal, diffusor in frosted and
tempered crystal
PHOTOGRAPHY
Foto Mosna

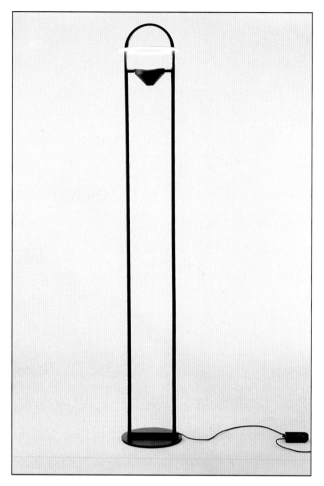

PRODUCT
"Circus" floor lamp
DESIGNER
Roberto Marcatti
CLIENT
Noto
PHOTOGRAPHY
Bitetto-Chimenti

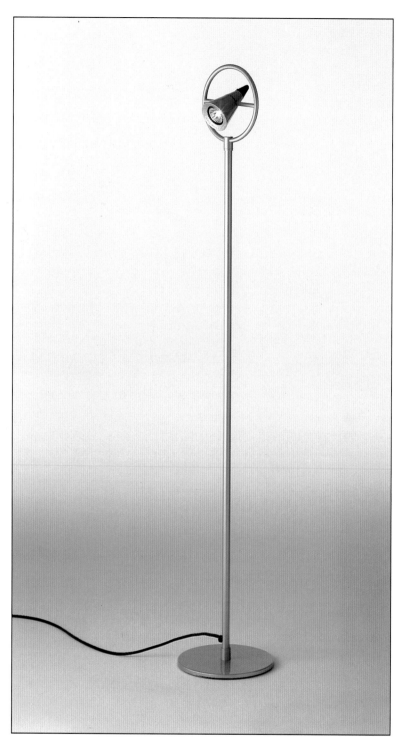

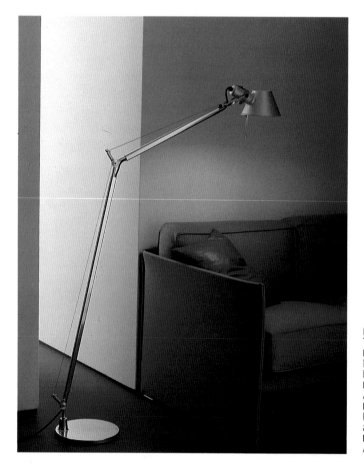

PRODUCT
"Speaker" floor and table lamp
DESIGNER
Davide Mercatali
CLIENT
Noto
PHOTOGRAPHY
Bitetto-Chimenti

PRODUCT
Tolomeo Reading Floor
DESIGNERS
Michele de Lucchi; Giancarlo
Fassina
MANUFACTURER
Artemide Inc.
DESCRIPTION
Floor reading lamp with
adjustable lamp
DISTRIBUTOR
Artemide Inc.

PRODUCT
"Fritz" Lamp
DESIGNERS
Perry A. King; Santiago Miranda
MANUFACTURER
Flos Inc.
CLIENT
Flos Inc.

PRODUCT
Ará
DESIGNER
Philippe Starck
MANUFACTURER
Flos Inc.
CLIENT
Flos Inc.

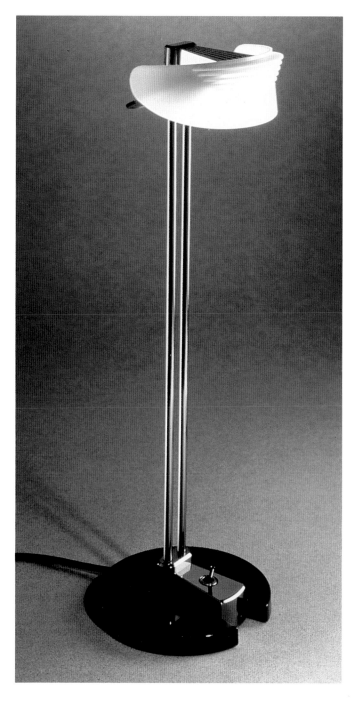

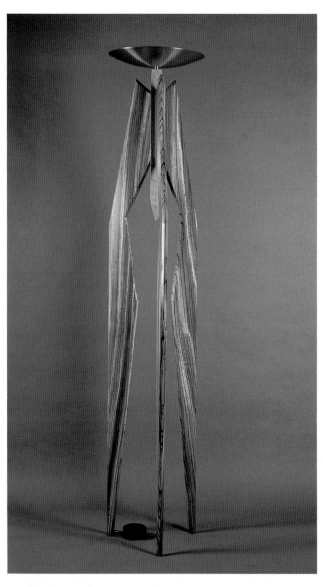

PRODUCT
Torchere Lamp
DESIGNER
Alan S. Kushner
DESCRIPTION
Constructed of zebra wood with a stained oak copper top and halogen lamp on Reistate.
PHOTOGRAPHY
Alice Sebrell

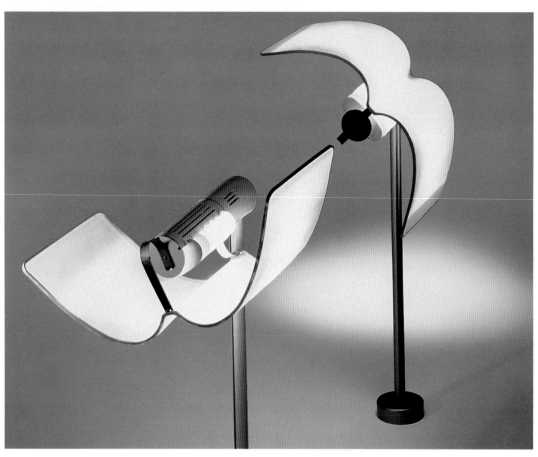

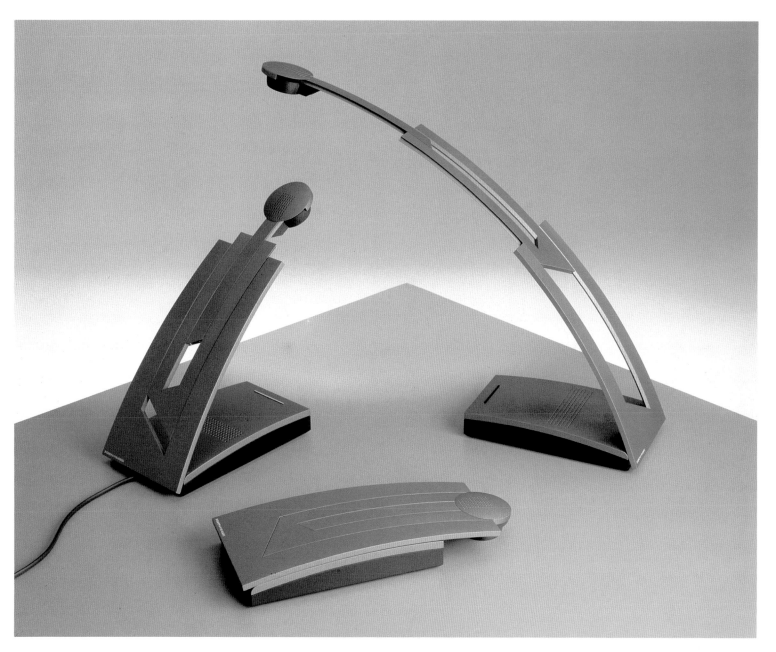

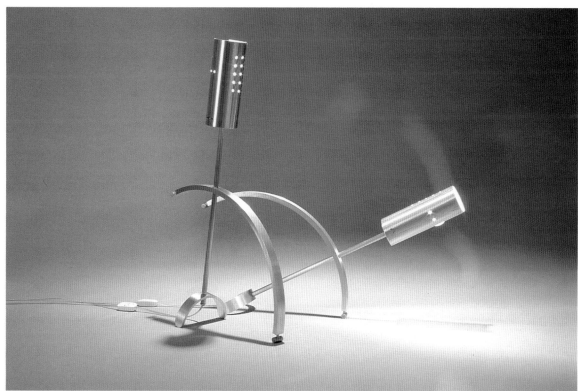

PRODUCT
Cartwheel Lamp
DESIGNER
James Evanson
FIRM
Lewis Dolin, Inc.
MANUFACTURER
James Evanson
DESCRIPTION
Made of copper and aluminum;
'stands' at almost 90 degrees and
'reclines' when flipped over
PHOTOGRAPHY
James Evanson

PRODUCT
"Dear Fausto" standing lamps
(full view)

PRODUCT
"Dear Fausto" standing lamp
(detail)
DESIGNER
Shigeru Uchida
FIRM
Studio 80
MANUFACTURER
Yamagiwa Co., Ltd.
DESCRIPTION
Deflector is of steel with a baked
melamine finish; body is of
Katsura wood with a clear
lacquer finish.
PHOTOGRAPHY
Nacása & Partners Inc.

126

PRODUCT
Constantin AP27635 Chandelier
DESIGNER
Andree Putman
MANUFACTURER
Baldinger Architectural
Lighting Inc.
DESCRIPTION
Has an integral downlight with
satin nickel finish and clear
frosted glass

PRODUCT
Suspenders
DESIGNER
Sonneman Design Group Inc.
FIRM
Sonneman Design Group Inc.
MANUFACTURER
George Kovacs Lighting, Inc.
DESCRIPTION
A modular halogen lighting system — mobilelike suspensions form luminous sculptures of balance and aesthetic tension. Materials include black metal bar stock, pressed ethed glass and PVC coated conductive rods.
CLIENT
George Kovacs Lighting, Inc.
PHOTOGRAPHY
Joseph Clementi

PRODUCT
Mikado System (detail)
DESIGNER
F.A. Porsche
MANUFACTURER
Artemide Litech Inc.
DESCRIPTION
Once the principle conductors are mounted in position, the rods may be arranged in a purely casual way or to compose geometrical figures to meet illumination requirements
DISTRIBUTOR
Artemide Inc.

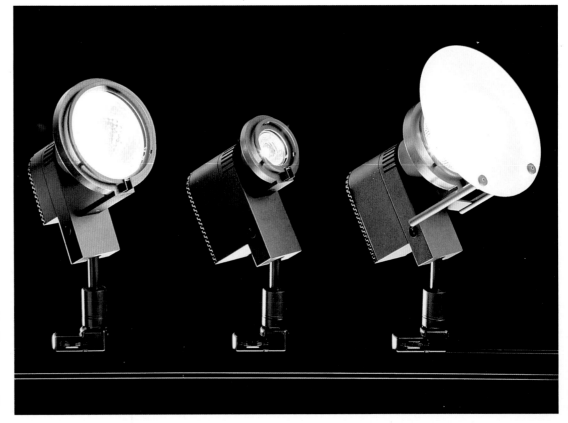

PRODUCT
Spots
DESIGNERS
Ninaber/Peters/Krouwel
FIRM
Ninaber/Peters/Krouwel
MANUFACTURER
Siemens
DESCRIPTION
Halogen spotlights
PHOTOGRAPHY
Ninaber/Peters/Krouwel
photostudio

PRODUCT
"Chip" Wall Sconce
DESIGNER
Piotr Sierakowski
FIRM
Koch + Lowy
MANUFACTURER
Koch + Lowy
PHOTOGRAPHY
Peter Weidlein

PRODUCT
"Luci Fair" Wall Sconce
DESIGNER
Philippe Starck
MANUFACTURER
Flos Inc.
CLIENT
Flos Inc.

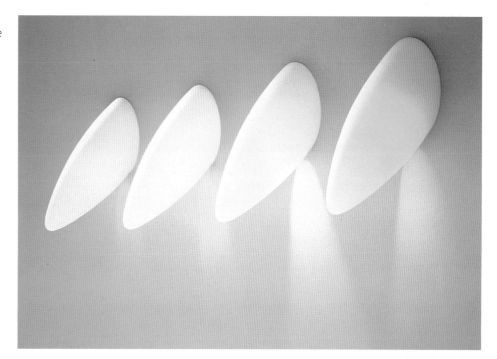

PRODUCT
Winged Victory
DESIGNER
David Baird
FIRM
Ziggurat
MANUFACTURER
Ziggurat
DESCRIPTION
Wall/floor lamp with galvanized
steel wings and a black staff
CLIENT
Ziggurat
PHOTOGRAPHY
Alan Linn

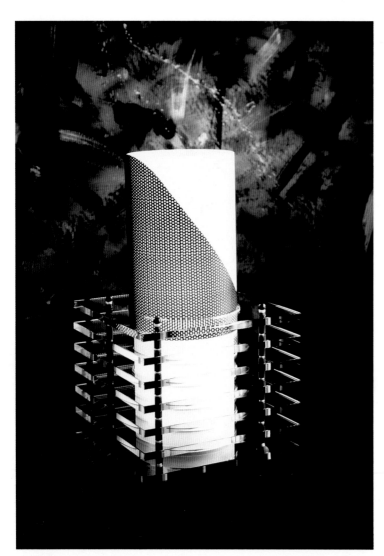

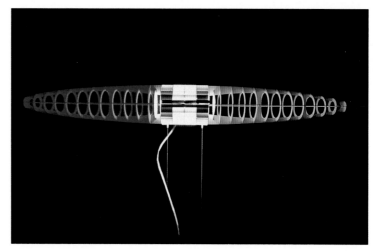

PRODUCT
Titania
DESIGNERS
Alberto Meda; Paolo Rizzatto
MANUFACTURER
Luce Plan
DESCRIPTION
Suspension lamp for direct and
reflected lighting. Adjustable
suspension cables are independent
of electrical connection which
permits different positioning of
the lamp in space; has five
interchangeable color filters
DISTRIBUTOR
Artemide Inc.

PRODUCT
Formosa
DESIGNERS
Roberto Marcatti, Alfonso Crotti
FIRM
Lavori in Corso
MANUFACTURER
Ar Far Studio Luce
DESCRIPTION
Table lamp in crystal or treated
copper, swivel screen in black
painted metal
PHOTOGRAPHY
Foto Mosna

PRODUCT
Conehead Sconce
DESIGNER
Frederic Schwartz
FIRM
Anderson/Schwartz Architects
MANUFACTURER
ASAP
DESCRIPTION
Sconces have silk shades.
PHOTOGRAPHY
Elliot Kaufman

PRODUCT
Y.M.D. IMONO floorstand "Arare"
DESIGNER
Takenobu Igarashi
FIRM
Igarashi Studio
MANUFACTURER
Yamasho Casting
DESCRIPTION
Created by using traditional cast
iron methods used in Yamagata
Prefecture for over 900 years
PHOTOGRAPHY
Masaru Meru
DISTRIBUTOR
Yamada Shomei Lighting Co.,
Y.M.D. Division

PRODUCT
Y.M.D. IMONO floorstand
"Kakiwari"
DESIGNER
Takenobu Igarashi
FIRM
Igarashi Studio
MANUFACTURER
Yamasho Casting
DESCRIPTION
Created by using traditional cast
iron methods used in Yamagata
Prefecture for over 900 years
PHOTOGRAPHY
Masaru Meru
DISTRIBUTOR
Yamada Shomei Lighting Co.,
Y.M.D. Division

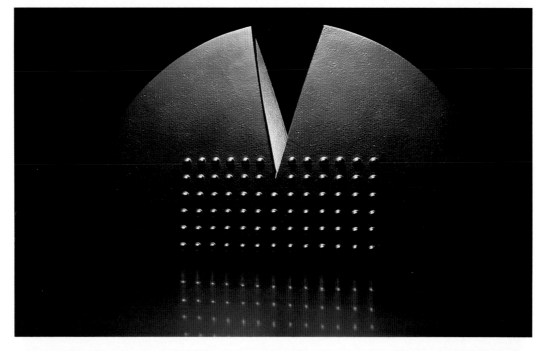

PRODUCT
Y.M.D. IMONO floorstand
"Danochi"
DESIGNER
Takenobu Igarashi
FIRM
Igarashi Studio
MANUFACTURER
Yamasho Casting
DESCRIPTION
Created by using traditional cast
iron methods used in Yamagata
Prefecture for over 900 years
PHOTOGRAPHY
Masaru Meru
DISTRIBUTOR
Yamada Shomei Lighting Co.,
Y.M.D. Division

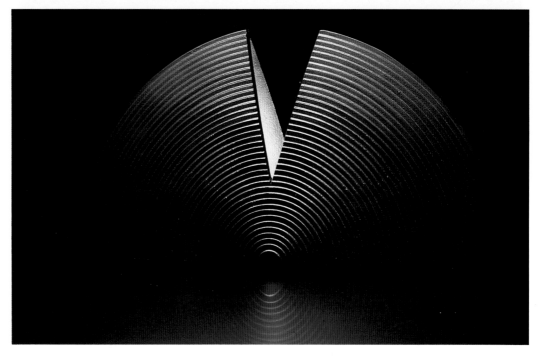

5 Electronics/ Appliances

PRODUCTS ARE THE "TOOLS" WE use to live our lives. In today's world they are becoming increasingly pervasive and the role they play in shaping our lives is profound indeed. Products have the ability to enhance our existence – if they are easy and enjoyable to use – or to hinder it. Everyone has experienced the frustration of using a product that has great potential which cannot be realized because it is difficult to use. Because products play such an important role in our lives, it is critical that they be well designed. Effective "tools" help each of us to realize our own potential.

The greatest challenge facing designers today involves helping industry to produce the highest quality products possible. Designers have an important contribution to make to this effort, and it encompasses more than just developing good designs. We have been developing good designs since the birth of the profession, but they have not always resulted in well-designed products. This is most often the case when the designer is not involved in the implementation of the design. The key lies in implementation.

The climate in industry has changed in a way which presents an opportunity to designers. Corporations are more frequently looking to outside resources

for product development services. And they would prefer a complete solution rather than a design concept or theoretical design. This provides the ideal situation for designers to exert greater influence on the quality of manufactured products.

I can think of no other group of people as well suited to filling this role as designers. We have the ideal mix of skills, knowledge and sensitivities needed to not only generate the ideas required, but to ensure that they are executed with the utmost care.

JIM PAGELLA
Program Manager of Industrial Design
GREGORY FOSSELLA DESIGN

Jim Pagella is currently the Program Manager of Industrial Design at Gregory Fossella Design, where he directs projects for clients in the medical instrumentation, computer and commercial equipment industries.

Prior to joining GFD, Mr. Pagella was a designer with NCR Corporation, responsible for the development of retail terminals and systems. A graduate of the University of Bridgeport, he has been recognized by IDSA and ID magazine for his work on products for Acoustic Research and Nova Biomedical.

PRODUCT
VICOM Prototype (Voice-Image
Communications Station Set)
(open view)
DESIGNER
Eric Chan
FIRM
ECCO
DESCRIPTION
Unique handset design offers on-
hook/off-hook dialing convenience.
Features include: voice message –
can be recorded in form of voice
or data to be stored / displayed /
printed out; has fax transmission
capabilities.
CLIENT
NYNEX
PHOTOGRAPHY
M. Ferrari / E. Chan

PRODUCT
Tsunami Telephone
DESIGNER
Mike Nuttall
FIRM
Matrix Product Design
CLIENT
Tsunami
PHOTOGRAPHY
John Long

PRODUCT
Legame cordless telephone
DESIGNER
Takenobu Igarashi
FIRM
Igarashi Studio
MANUFACTURER
Nittsuko
DESCRIPTION
Both units function as a phone with hardly any frequency inter-ferences; comes in black and silver.
PHOTOGRAPHY
Dale Berman
DISTRIBUTOR
FOLMER Corp.

PRODUCT
EC II Phone
DESIGNER
Eric Chan
FIRM
ECCO
DESCRIPTION
Two-line telephone with ergonom-ically-profiled handset
PHOTOGRAPHY
M. Ferrari/E. Chan
DISTRIBUTION
Becker Inc.

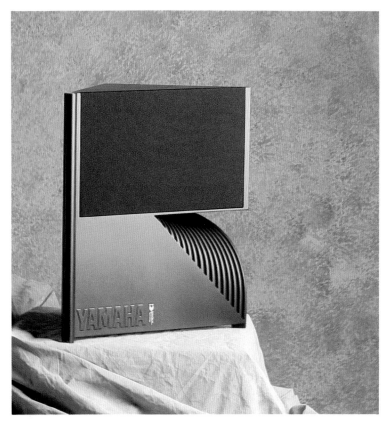

PRODUCT
Speakers
DESIGNER
Vlad Müller MA
FIRM
Müller Ullmann
DESCRIPTION
Two-part injection molded
enclosure speakers
PHOTOGRAPHY
Yuri Dojc

PRODUCT
Epicure Home Loudspeakers
DESIGNER
Daniel Ashcraft, Scott Shimatsu,
Scott Shulman
FIRM
Ashcraft Design
CLIENT
Epicure

PRODUCT
Control Micro Loudspeaker
DESIGNERS
Daniel Ashcraft, Scott Shimatsu
FIRM
Ashcraft Design
CLIENT
JBL International
PHOTOGRAPHY
Herve Grison

PRODUCT
Audio Speaker AST-X
DESIGNER
Hiroaki Kozu
FIRM
Kozu Design
MANUFACTURER
Yamaha Corp.
DESCRIPTION
High-performance compact audio
speaker with new ceramics for
speaker enclosure
CLIENT
Yamaha Corp.
AWARD
Grand Prize, 1989 Yamaha AST
System International Design
Competition

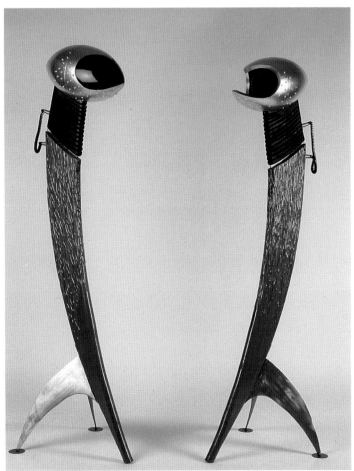

PRODUCT
"Iliad" 1989 Speakers
DESIGNER
Alex Locadia
MANUFACTURER
Alex Locadia and Art et Industrie

PRODUCT
HD 7500 CD Player
DESIGNER
Daniel Ashcraft, Scott Shimatsu
FIRM
Ashcraft Design
CLIENT
Harman/Kardon
PHOTOGRAPHY
Herve Grison

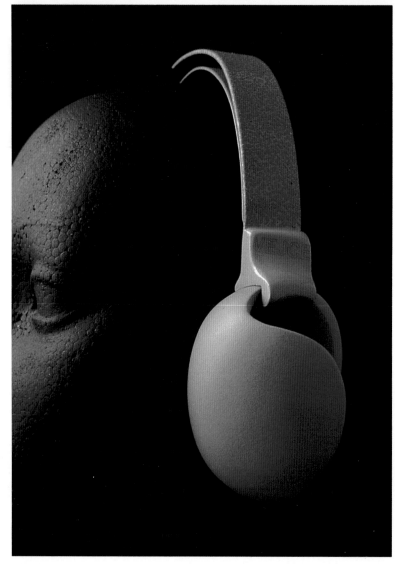

PRODUCT
Ear Cover/Sound Deflector
DESIGNER
Olga Ullman MA
FIRM
Müller Ullmann Industrial Design
DESCRIPTION
Injection-molded, parabolic ear
shells eliminate noise created by
ear turbulence while bicycling,
jogging or skiing.
CLIENT
ACW Technology
PHOTOGRAPHY
Vlad Müller

PRODUCT
Interactive Video Hand-held
Terminal
DESIGNER
Michael Barry
FIRM
GVO, Inc.
DESCRIPTION
Used with special television
broadcast programming, this
device allows access to a world of
real time events, game shows,
home shopping, banking,
electronic polling, market
research and educational media
CLIENT
Interactive Network
PHOTOGRAPHY
Mark Gottlieb

PRODUCT
Mastermind Universal Remote
DESIGNERS
Daniel Ashcraft, Hiro Teranishi
FIRM
Ashcraft Design
CLIENT
Harman/Kardon

PRODUCT
Voice Computer
DESIGNER
Jeff Smith
FIRM
Lunar Design Inc.
MANUFACTURER
Talknology
CLIENT
Talknology
PHOTOGRAPHY
Rick English

PRODUCT
ER-500 Cash Register
DESIGNER
Sharp Corporate Design Center,
Tamotsu Yoshikawa
FIRM
Sharp Corporation
DESCRIPTION
For domestic market

PRODUCT
Double Plus calculator
DESIGNER
Noel Zeller
FIRM
Zelco Industries Inc.
MANUFACTURER
Zelco Industries Inc.
DESCRIPTION
There are two models: one for
right-hand users and one for left-
hand users. There are also two
"plus" keys: for high-speed
calculations and for desktop use.

PRODUCT
Vivitar X-300
DESIGNER
Fernando Pardo, James Grove,
Steven Shull
FIRM
Vivitar Corp.
DESCRIPTION
Auto-focus 35mm compact
camera
CLIENT
Vivitar Corp.
PHOTOGRAPHY
Steve Breitborde

PRODUCT
Steadicam® / Jr.
DESIGNERS
Gerald Skulley, Robert J. Hayes
FIRM
Fitch RichardsonSmith
MANUFACTURER
Cinema Products Corp.
DESCRIPTION
Provides the means to create
shake-free, fluid action scenes for
users of 8mm and VHS-C cam-
corders weighing under 3 pounds.
CLIENT
Cinema Products Corp.
PHOTOGRAPHY
Courtesy of Joel Lipton

PRODUCT
Thumper
DESIGNER
Noel Zeller
FIRM
Zelco Industries Inc.
MANUFACTURER
Zelco Industries Inc.
DESCRIPTION
Alarm clock

PRODUCT
Cord Central
DESIGNER
Paul Bradley
FIRM
Matrix Product Design
MANUFACTURER
Homestar
DESCRIPTION
Auto-retractable extension cords
CLIENT
Homestar International
PHOTOGRAPHY
Rick English

PRODUCT
Electric Egg Boiler
DESIGNERS
Marco Susani, Richard Eisermann
FIRM
Sottsass Associati
DESCRIPTION
Made of ceramic materials and
tempered glass, egg boiler acts as
serving tray when inverted.
Measuring device and egg piercer
are integrated.
CLIENT
Bodum

PRODUCT
The Breakfast Totem
DESIGNERS
Marco Susani, Richard Eisermann
FIRM
Sottsass Associati
DESCRIPTION
Juice squeezer stacks on top of
egg boiler, saving counter space,
creating a "breakfast totem."
CLIENT
Bodum

PRODUCT
Bodum Juice Squeezer
DESIGNERS
Marco Susani, Richard Eisermann
FIRM
Sottsass Associati
DESCRIPTION
Ceramic materials and circular
shapes have a universal appeal
that also enhances function
CLIENT
Bodum

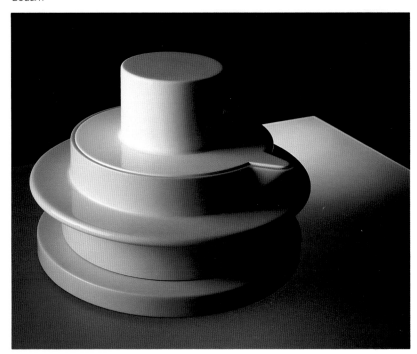

PRODUCT
The Expert
DESIGNER
Marlan Polhemus
FIRM
Goldsmith Yamasaki Specht
MANUFACTURER
Mr. Coffee
DESCRIPTION
Electronic programmable brewing cycle coffee maker with 24-hour digital timer; has pause 'n serve feature
CLIENT
Mr. Coffee
PHOTOGRAPHY
Ed Nagel

PRODUCT
Tower Toaster
DESIGNER
GYS Staff
FIRM
Goldsmith Yamasaki Specht
MANUFACTURER
Concept
DESCRIPTION
Toaster opens to allow bread to pass through while toasting
CLIENT
GYS
PHOTOGRAPHY
GYS: Dan Karp

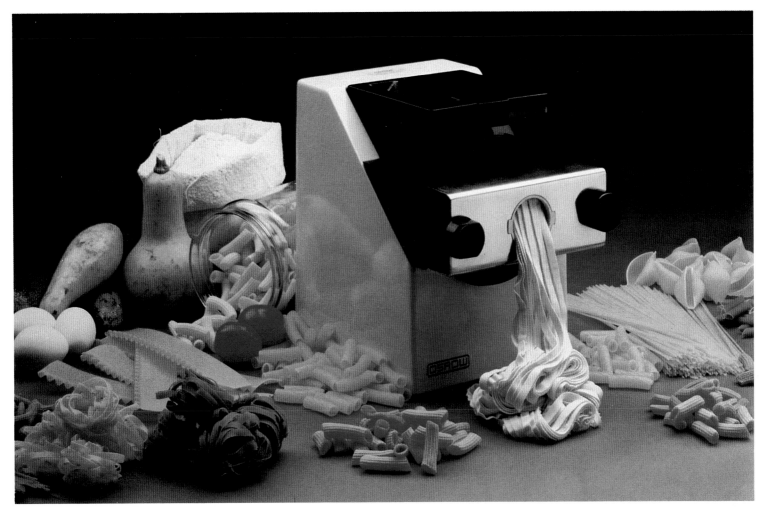

PRODUCT
Pasta Express
DESIGNER
Gordon Randall Perry
FIRM
Gordon Randall Perry Design Inc.
MANUFACTURER
Creative Technologies
DESCRIPTION
It kneads, extrudes, and dries the dough in a variety of shapes and flavors
CLIENT
Osron Products (original manufacturer)
AWARD
Participant in *Design in America* show in USSR

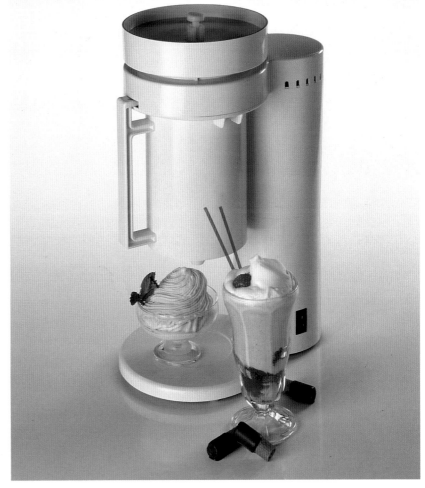

PRODUCT
Ice Cream Machine
DESIGNER
Gordon Randall Perry
FIRM
Gordon Randall Perry Design Inc.
MANUFACTURER
Creative Technologies
DESCRIPTION
Makes fast desserts out of almost any sweet liquid
CLIENT
Creative Technologies (TAKKA)

PRODUCT
Module-Aire™ Kitchen
MANUFACTURER
ABBAKA
DESCRIPTION
Four individual interconnecting sections allows mixing and matching of decorative metal or enamel finishes
CLIENT
ABBAKA
PHOTOGRAPHY
Musilek Photography

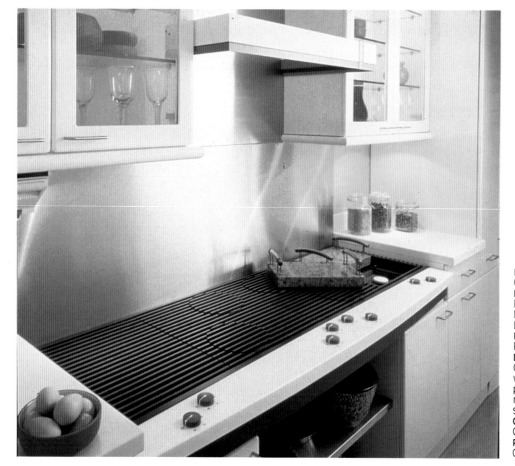

PRODUCT
Cooktop (concept)
DESIGNERS
Peter A. Koloski, Robert W. Mervar, Kenneth M. Brazell, Chuck Leinbach
FIRM
Fitch RichardsonSmith
DESCRIPTION
Gas burner stove top, gas grill with exhaust fan and a set of portable gas heated granite "hot rocks" were integrated into a single unit
CLIENT
G.E. Plastics
PHOTOGRAPHY
Courtesy of G.E. Plastics

PRODUCT
Multi-function oven w/electric
baking/pizza stone
FIRM
Gaggenau USA Corp.
PHOTOGRAPHY
Claus Froh

PRODUCT
Plastic bottle shredder
(proposed product)
DESIGNER
Mark Steiner
FIRM
Mark Steiner Design
DESCRIPTION
For home use

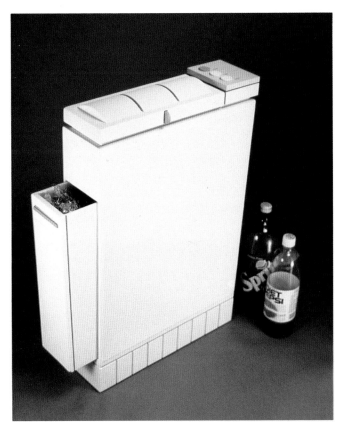

PRODUCT
"All In One" Stove
DESIGNER
James Hong
MANUFACTURER
James Hong and Art et Industrie
DESCRIPTION
Lacquer, steel, slate and aluminum

151

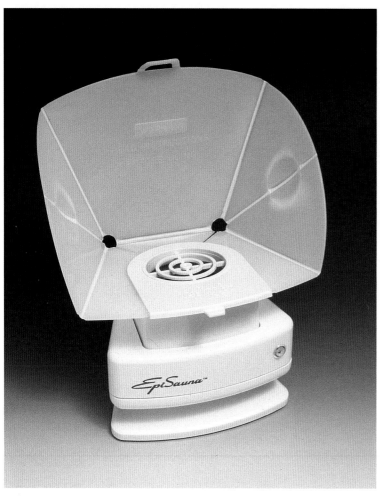

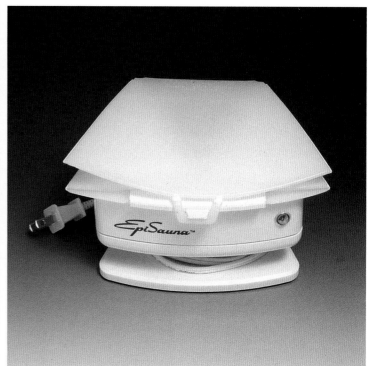

PRODUCT
Facial Sauna (closed view)

PRODUCT
Facial Sauna (open view)
DESIGNERS
Hiroaki Kozu, Michael Young
FIRM
Michael W. Young Associates, Inc.
CLIENT
Bernhard Industries, Inc. (later
sold to Epi Products U.S.A. Inc.)

PRODUCT
Curl Dazzler
DESIGNERS
James Howard, Acey Tzue
FIRM
Howard Design
MANUFACTURER
Conair Corp.
PHOTOGRAPHY
Tom de Guercio

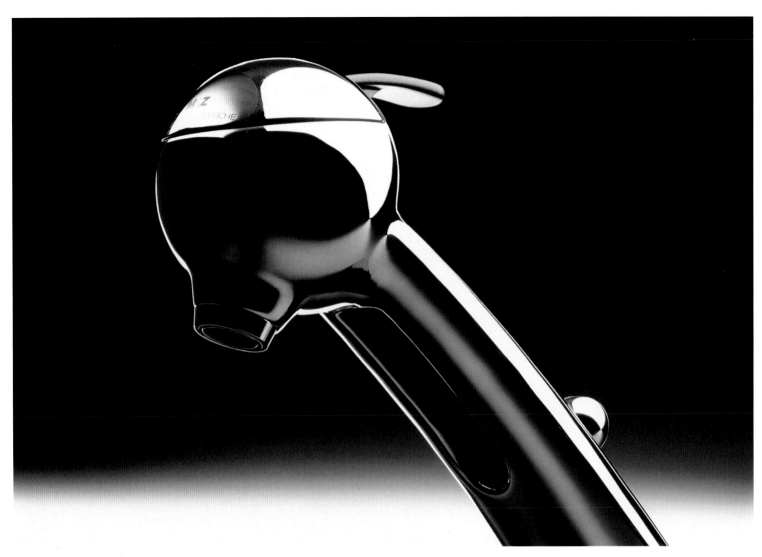

PRODUCT
Eclisse
DESIGNER
F.A. Porsche
FIRM
Porsche Design Gmb H
DESCRIPTION
Bathroom mixer: hot and cold water are brought together in a mixing chamber
DISTRIBUTOR
M & Z, Italy

PRODUCT
Loop Handle Faucet
DESIGNERS
Sam Disney, Andrew Alger
FIRM
Goldsmith Yamasaki Specht
MANUFACTURER
Eljer
DESCRIPTION
Epoxy-finish brass bathroom single-lever faucet
CLIENT
Eljer Plumbingware
PHOTOGRAPHY
Eljer

6 Recreational Equipment

THE PRODUCTS WE SELECT GREATLY reflect our taste and attitudes. I believe if we are offered increasingly better products, more environmentally friendly perhaps, then our perceptions may be changed for the better. The converse, of course, also applies.

Products are for people. Products answer needs, increase happiness or alleviate suffering.

I strive to design and produce products that look right and perform well. Looking right most often means looking beautiful, but sometimes products need to appear 'macho' or 'cuddly' or, if the brief is to design a gargoyle, down-right ugly.

A superb blend of aesthetics, usefulness and ease of use at a marketable price are, to my mind, the essential attributes of the perfect product. Personally, I am most interested in the design and production of fast moving consumer durables.

Due to competitive market forces, many products get better and better. Automobiles, for instance, break-down less often. Efficient writing instruments and disposable razors are now available to wide numbers of people at a fraction of their cost ten years ago.

I suspect that in the '90s, the product designer's job will become more difficult. Fashionable shapes will proliferate, and these are more demanding for the industrial designer than the straight lines which may have been acceptable in the '80s.

Competition is also heating up, especially in the area of product aesthetics, as more and more profit-conscious companies seek a market-winning edge with that 'right look'.

IAIN SINCLAIR
Iain Sinclair Design

Iain Sinclair brings almost 30 years of design experience to Iain Sinclair Ltd., one of England's leading design firms. Due to his diverse creative background, he has been commissioned for over 120 consumer-oriented products, including furniture, lighting, games, chess computers and numerous electronic-based hardware.

Mr. Sinclair, who has been featured at New York's Museum of Modern Art, is a Professor of Industrial Design at the Royal College of Art in London. Besides writing for design journals, he occasionally presents his expertise at lectures and seminars. From 1977 until his resignation in 1987, he was a Fellow of the Society of Industrial Artists and Designers.

He resides in Cambridge with his wife and three children.

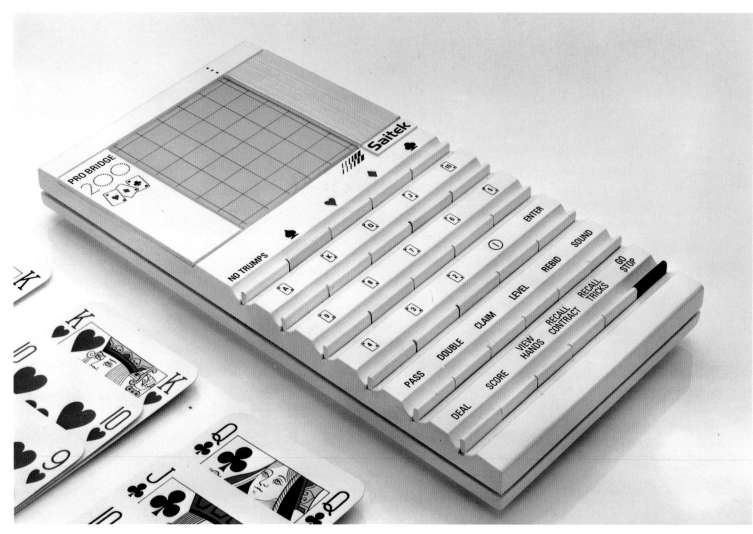

PRODUCT
Pro Bridge 200
DESIGNER
Iain Sinclair
FIRM
Iain Sinclair Design
DESCRIPTION
Hand-held bridge computer of
injection-molded plastic
CLIENT
Saitek Ltd., Hong Kong

PRODUCT
Kasparov Simulatano Chess
Computer
DESIGNER
Iain Sinclair
FIRM
Iain Sinclair Design
DESCRIPTION
Patented LCD chessboard of
injection-molded plastic.
ELO rating of 2200.
CLIENT
Saitek Ltd., Hong Kong

PRODUCT
Playframe
DESIGNER
Karen Hewitt
FIRM
Learning Materials Workshop, Inc.
MANUFACTURER
Learning Materials Workshop, Inc.
DESCRIPTION
104-piece playset constructed
from hardware and non-toxic
materials
PHOTOGRAPHY
Ken Burris
AWARD
1990 Parent's Choice Foundation
Award

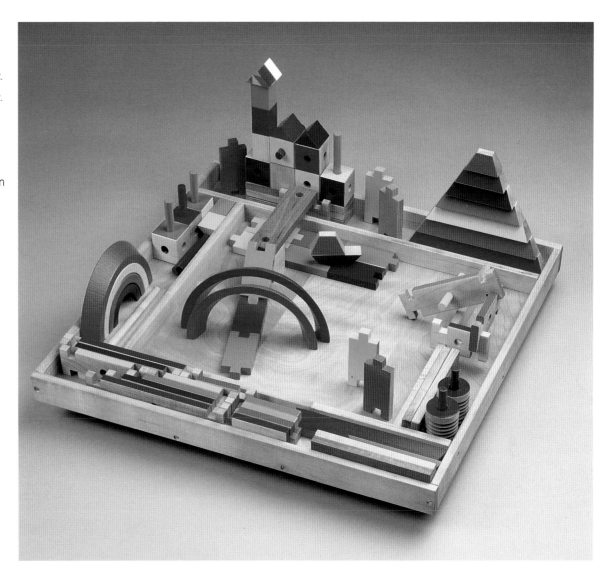

PRODUCT
Chess Set
DESIGNER
Kurt B. Delbanco
FIRM
Delbanco Arts
PHOTOGRAPHY
John Schwartz

PRODUCT
To Go! Products
DESIGNERS
Khodi Feiz, Kaz Amemiya
FIRM
Texas Instrument Corporate
Design Center
DESCRIPTION
A line of electronic educational
products for children ages 4-8
CLIENT
Texas Instruments

PRODUCT
Super Speak & Spell
DESIGNER
Khodi Feiz
FIRM
Texas Instrument Corporate
Design Center
DESCRIPTION
A line of electronic, educational
products for children ages 6-12,
teaching spelling, vocabulary and
math
CLIENT
Texas Instruments

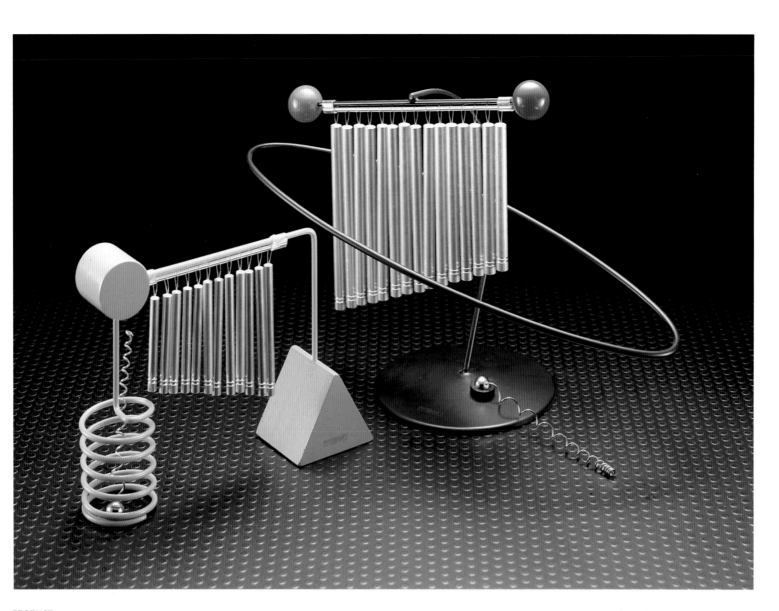

PRODUCT
Chimes
FIRM
Canetti Design Group
MANUFACTURER
Canetti Inc.
CLIENT
Canetti Inc.
PHOTOGRAPHY
Color Track

PRODUCT
Feet Good
FIRM
Canetti Design Group
MANUFACTURER
Twinbird
DESCRIPTION
Solid plastic foot massager with
three different textures
CLIENT
Canetti Inc.
PHOTOGRAPHY
Color Track

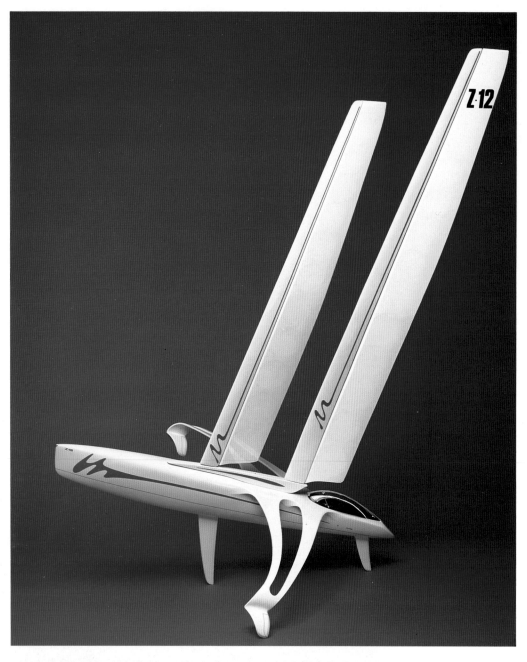

PRODUCT
Sailboat
DESIGNER
Ricardo Salinas
FIRM
Ricardo Salinas Industrial Design
DESCRIPTION
High-performance sailboat with rigid 12-meter sails
PHOTOGRAPHY
Manuel Hieblum

PRODUCT
O'Neill high performance wetsuit "The Animal"
DESIGNER
Vent Design Associates
FIRM
Vent Design Associates
MANUFACTURER
O'Neill, Inc.
DESCRIPTION
A high-end wetsuit cut to create a truly contoured envelope around the body featuring molded expansion areas for easier mobility.
CLIENT
O'Neill, Inc.
PHOTOGRAPHY
Rick English

PRODUCT
Bowflex
DESIGNERS
Marlan Polhemus, Scott
Vermillion, Paul Specht
FIRM
Goldsmith Yamasaki Specht
MANUFACTURER
Schwinn
DESCRIPTION
Progressive resistance exercise
machine
CLIENT
Schwinn
PHOTOGRAPHY
Photoworks

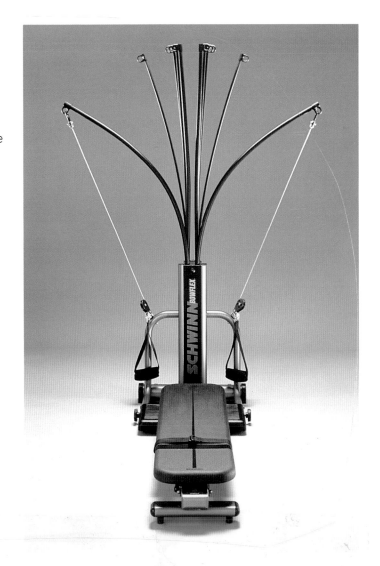

PRODUCT
Exerciset
FIRM
Canetti Design Group
MANUFACTURER
Canetti, Inc.
DESCRIPTION
Set of five different pieces of
exercise equipment
CLIENT
Canetti, Inc.
PHOTOGRAPHY
Color Track

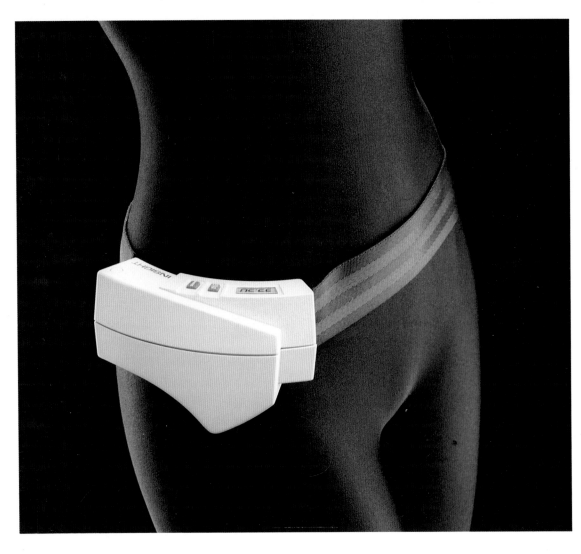

PRODUCT
Accelerator
DESIGNERS
Nick Barker; Jeff Brown
FIRM
Technology Design
DESCRIPTION
Measures skiiers', runners', and bicyclists' speed and distance using microwave
CLIENT
Insight Corp.
PHOTOGRAPHY
Jeff Curtis Photography

PRODUCT
Veloce Racing Helmet
PROJECT MANAGEMENT
Bob Read; Bill Corliss
FIRM
Tres Design Group
PROJECT ENGINEER
M.P.A., Turin, Italy
DESCRIPTION
Aerodynamic Lexan polycarbonate helmet
CLIENT
Schwinn Bicycle Co./Paramount Brand
PHOTOGRAPHY
John Payne

PRODUCT
Recreational Headwear
DESIGNER
David Wiener Ventures
FIRM
David Weiner Designs
CLIENT
Kombi Ltd.

PRODUCT
"Carbon" Cycling Footwear
PROJECT MANAGEMENT
Shinpei Okajima; Satoshi Watanabe
FIRM
Tres Design Group
PROJECT ENGINEER
Shimano, Japan
DESCRIPTION
Pearlized leather and high-strength ballistic mesh footwear
CLIENT
Shimano Industrial Co., Ltd.
PHOTOGRAPHY
John Payne

PRODUCT
Team Issue Racing Shoe
PROJECT MANAGEMENT
Bob Read; Bill Corliss
FIRM
Tres Design Group
PROJECT ENGINEER
Duigi, Italy
DESCRIPTION
Cycling shoe with inner and outer Velcro closure system eliminates pressure points; constructed of leather and nylon mesh
CLIENT
Schwinn Bicycle Co./Paramount Brand
PHOTOGRAPHY
John Payne

PRODUCT
Team Issue "Hard Core" Racing Helmet
PROJECT MANAGEMENT
Bob Read; Bill Corliss
FIRM
Tres Design Group
PROJECT ENGINEER
M.P.A., Italy
DESCRIPTION
Ultralight professional helmet fabricated in high-impact absorbing expanded EPS foam bonded to a microshell all wrapped around a high-density polypropetene core
CLIENT
Schwinn Bicycle Co./Paramount Brand
PHOTOGRAPHY
John Payne

7 Office/Medical Equipment

I BELIEVE THE PRODUCTS WE design and use help shape us and our society. Just consider telephones and computers. Because we as designers shape the products, we must work in a responsible way and at the same time stretch our creativity to find better solutions. For example, we must better integrate high technology products into our society, because now they often tend to alienate people. We can solve that problem by making them fundamentally understandable, easy to control and easy to repair. It's something we must do.

My goal when designing is to create a product that expresses its function and manufacture in a beautiful and clear way. I also like to turn things inside out or make new combinations, because sometimes the new arrangement works better and is more exciting than the old. To me, mass production is not just a series of design problems to be solved, it's an artist's palette that allows us to express our concepts by the tens of thousands in a unique medium. I'd rather design a parting line to be obvious and express tension in the form, than try to mask it. I espouse a kind of surprise industrial aesthetic which lets both my thinking and the beauty of the manufacturing process shine through.

Regrettably, I think the design profession has become too often the exclusive tool for marketing, which these days usually means we engage in a kind of focused bottom line thinking. And that is hurting our economy. In the future I see designers flexing their muscles more around social responsibility and ecological issues. I also see designers becoming more entrepreneurial, not only creating products, but having the products manufactured and marketed under their direction. I believe we have more strength than we think we have; strength that comes from being innovators and problem solvers. We're going to be busy in the days ahead.

Gordon Perry is an individual devoted to design. As head of his New York-based industrial design consulting firm, he designs and develops contract furniture, medical, scientific, lighting, office, houseware and architectural products. He is also a part-time professor of design at Pratt Institute, where he received a Master's Degree and a Bachelor of Industrial Design degree.

A former elected officer of The Industrial Designers Society of America, Mr. Perry has been a guest lecturer for the American Management Association, the National Art Education Association, Interiors magazine and the IDSA. His designs and viewpoints have appeared on the pages of Attenzione, Chicago Tribune, Industrial Design, Interior Design, Interiors, Metropolitan Home, Modern Office Technology and Product Design 2. The owner of more than a dozen patents, he is the recipient of awards from IBD, Industrial Design magazine and the American Iron and Steel Institute.

GORDON PERRY
Gordon Perry Design

PRODUCT
DIACAM, Medical Diagnostic
Nuclear Camera System
DESIGNER
Henry J. Rahn
DESCRIPTION
Only completely-balanced nuclear
camera available
CLIENT
SIEMENS
PHOTOGRAPHY
Tom Balla

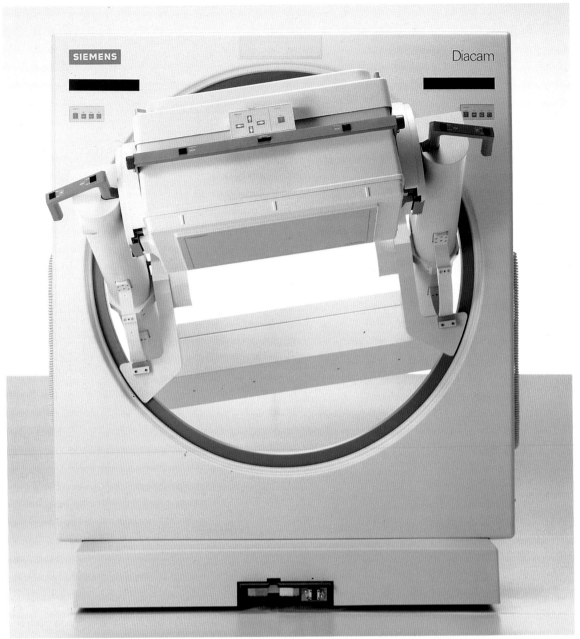

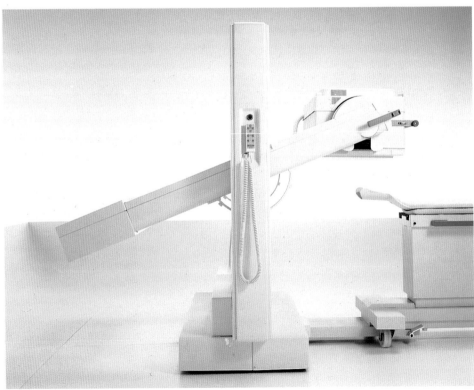

PRODUCT
Incubator
DESIGNERS
Joel Carpenter, Tom Burchard,
Michael Dann
FIRM
The Design Works
MANUFACTURER
Air Shields Vickers
DESCRIPTION
Pediatric intensive care unit
CLIENT
Air Shields Vickers
PHOTOGRAPHY
Kevin O'Riely

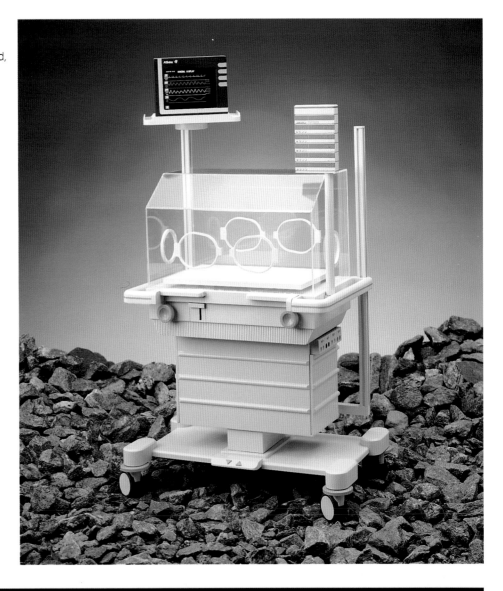

PRODUCT
Lab Scanning Device
DESIGNER
Jeff Smith
FIRM
Lunar Design Inc.
MANUFACTURER
Molecular Dynamics
CLIENT
Molecular Dynamics
PHOTOGRAPHY
Rick English

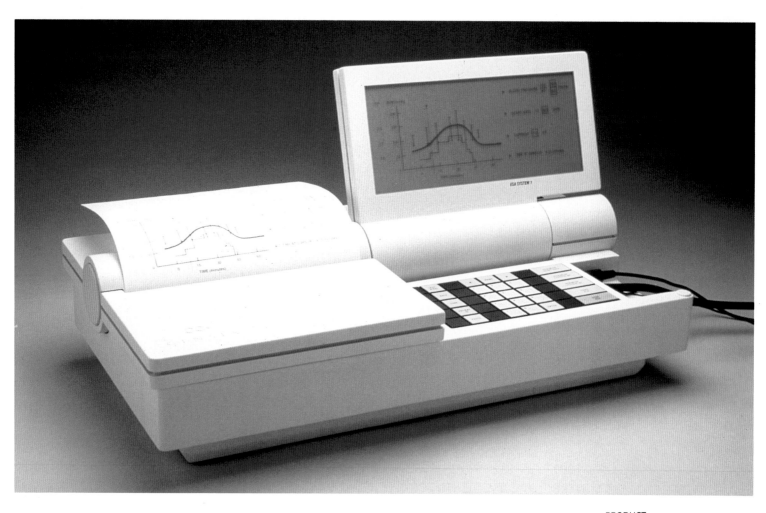

PRODUCT
ESA System
DESIGNER
Ron Boeder IDSA
FIRM
Boeder Design
MANUFACTURER
Ampex Corp.
DESCRIPTION
ESA system for the diagnosis and
treatment of heart disease
CLIENT
Ampex Corp.
PHOTOGRAPHY
Amanda Hatherly

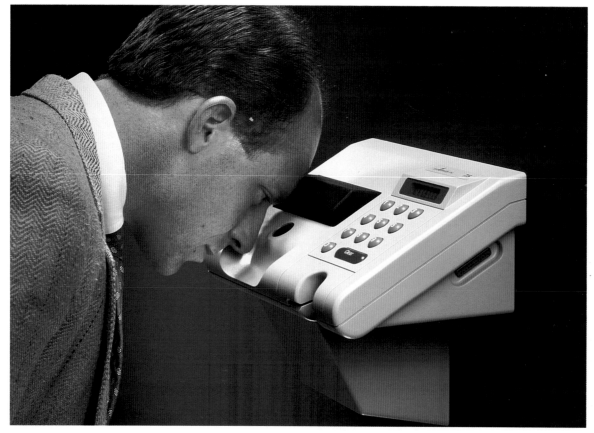

PRODUCT
7.5 Eyedentify (Retina Scanner)
DESIGNERS
LeRoy J. LaCelle, Anthony Grasso,
David Littrell
FIRM
Designhaus Inc.
DESCRIPTION
Individual identification device
which scans unique retinal blood
vessel patterns via infrared beam
with an error rate of one in a
million
CLIENT
Eyedentify
PHOTOGRAPHY
Dave Robinson

PRODUCT
Hall Versipower Dual Power
Orthopaedic Surgical Instrument
DESIGNERS
Bill Bartlett, Gerry Mielcarek
FIRM
Bartlett Design Associates, Inc.
DESCRIPTION
Battery-powered or corded, light-
weight unit provides greater
maneuverability, has a variety of
attachments and simplifies
orthroscopy
PHOTOGRAPHY
Mehosh

PRODUCT
RX-2000 Operator's X-Ray Console
DESIGNERS
John Amber, Jay Wilson
FIRM
GVO, Inc.
DESCRIPTION
Cast aluminum desktop operator's
console for controlling large X-ray
equipment used in non-destructive
testing
CLIENT
Varian Corp.
PHOTOGRAPHY
Mark Gottlieb

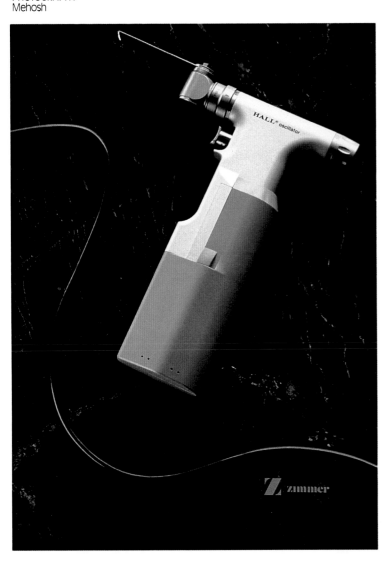

171

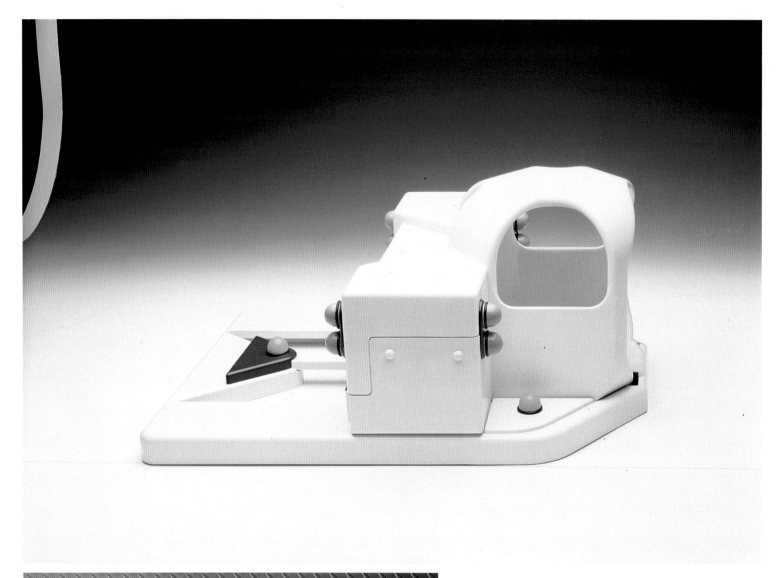

(side view)

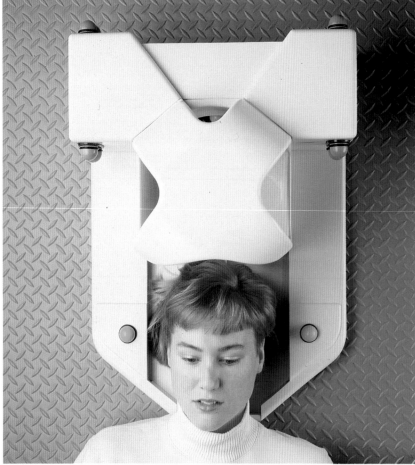

PRODUCT
Resonex VFQ
DESIGNER
John Amber
FIRM
GVO, Inc.
DESCRIPTION
Allows for high-resolution
scanning of the patient's head
for medical diagnosis
CLIENT
Resonex
PHOTOGRAPHY
Mark Gottlieb

PRODUCT
Traction Chair (full side view)
DESIGNER
Vincent L. Haley
FIRM
V.H. Designs
DESCRIPTION
Relieves pain in lumbar or cervical region by applying traction
CLIENT
V. H. Designs
PHOTOGRAPHY
Mark Steele
AWARD
Second Place Winner, Unysis Design Corp.

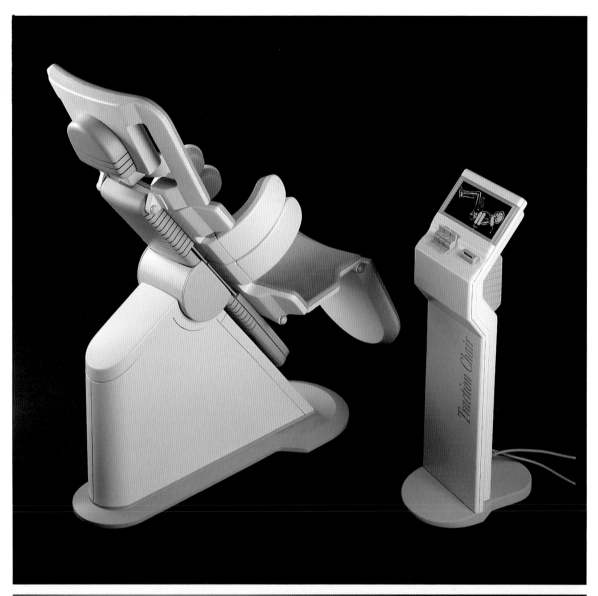

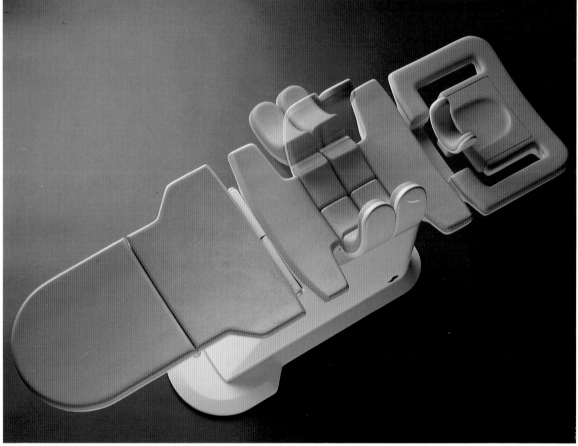

(aerial view)

PRODUCT
GCI 501
DESIGNER
Ravi Sawhney
FIRM
RKS Design
MANUFACTURER
GC Design
DESCRIPTION
Sounds an alarm when ambient
oxygen level fall dangerously low
PHOTOGRAPHY
Ravi Sawhney, Donald Brown

PRODUCT
Medical Argon Laser
DESIGNER
Jeff Smith
FIRM
Lunar Design Inc.
MANUFACTURER
Coherent Medical
CLIENT
Coherent Medical
PHOTOGRAPHY
Rick English

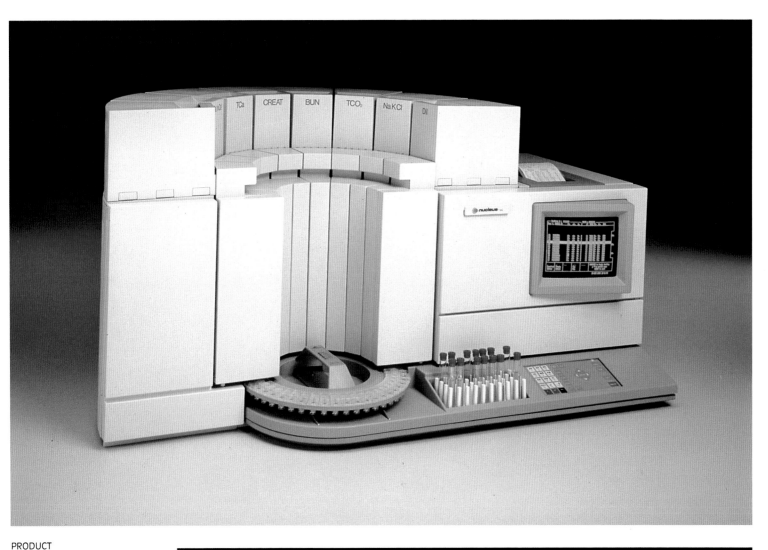

PRODUCT
Nova Nucleus Blood Analyzer
DESIGNERS
Jim Pagella, Richard Randall,
Gregory Kenny, Ben White
FIRM
Gregory Fossella Design, Inc., a
division of the McCalla/Lackey
Corp.
MANUFACTURER
Nova Biomedical
CLIENT
Nova Biomedical
AWARDS
1990 IDEA Gold Award for Clinical
and Diagnostical Products; ID
Magazine Design Review Selection

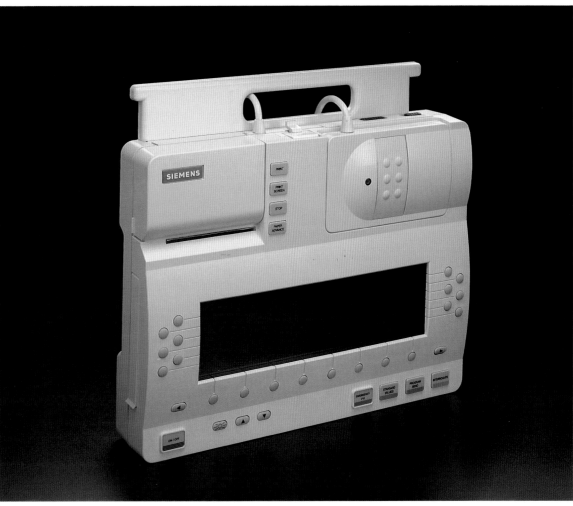

PRODUCT
Pacesetter — Siemens Pacemaker
Programmer
DESIGNER
Bill Bartlett, David Hines,
Kevin Clay
FIRM
Bartlett Design Associates, Inc.
DESCRIPTION
Units are programmed to specific
heart rates through a small
handheld telemetry device held
against the patient's chest
PHOTOGRAPHY
Mehosh
AWARD
Third Place ID Society Award for
Industrial Design

PRODUCT
Autodiagnostic Bone Growth
Stimulator
DESIGNER
Ravi Sawhney, Dennis Wasserman,
Donald Brown
FIRM
RKS Design
MANUFACTURER
Ace Medical
DESCRIPTION
Expedites progress of severely
broken bones
PHOTOGRAPHY
Don Brown, Dan Hutchings

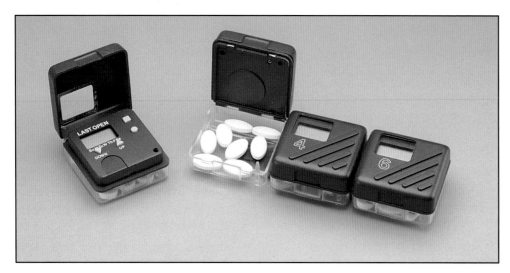

PRODUCT
Multiprescript
DESIGNER
Ralph De Vito, Tony Di Gangi,
James Howard
FIRM
Howard Design
MANUFACTURER
Wheaton Medical
DESCRIPTION
Pill box with built-in digital timer
PHOTOGRAPHY
Tom del Guercio

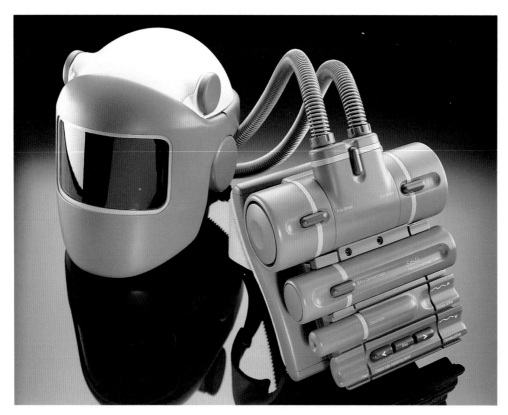

PRODUCT
SafeAir Personal Environment
System
DESIGNER
Vincent L. Haley
FIRM
V. H. Designs
DESCRIPTION
Protects welders and foundry
workers from hazardous
environments
PHOTOGRAPHY
Mark Steele/Steve Trank

PRODUCT
Porta Support
DESIGNER
Ravi Sawhney, Dennis Wasserman,
Donald Brown, Hiro Teranishi
FIRM
RKS Design
MANUFACTURER
Sommers Concept
PHOTOGRAPHY
Ravi Sawney, Donald Brown,
Dan Hutchings
AWARD
On The Edge 1990 Selection

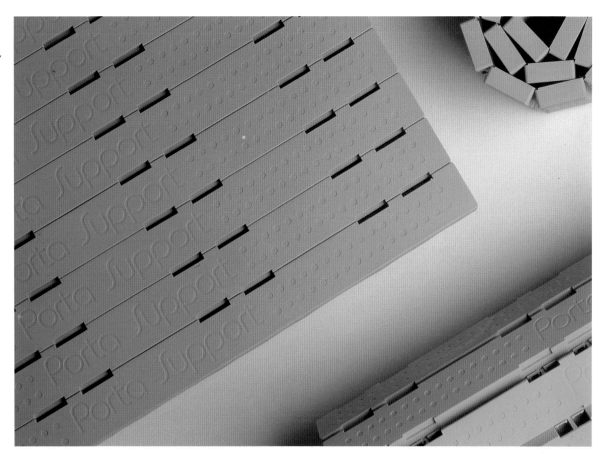

PRODUCT
Finally Firm
DESIGNERS
Ravi Sawhney, Donald Brown, Dan
Hutchings, Hiro Teranishi
FIRM
RKS Design
MANUFACTURER
Selvac Corp.
PHOTOGRAPHY
Ravi Sawhney, Donald Brown, Dan
Hutchings

PRODUCT
Sation 1000 Water Purifier
DESIGNER
Joan Sunyol
FIRM
Via Design S.A.
MANUFACTURER
Sation S.A.
CLIENT
Sation S.A.
PHOTOGRAPHY
Vandellos
AWARD
1989 Chamber of Commerce
Award

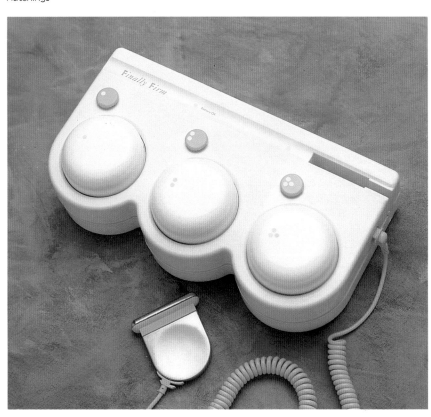

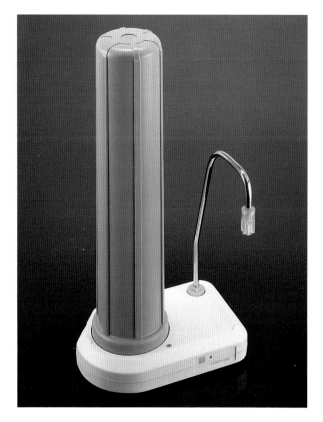

PRODUCT
Portable Office (Concept)
DESIGNER
Eric Chan
FIRM
ECCO
DESCRIPTION
A portable office that permits
people to work/communicate at
any time and place: combines
computer, telephone, fax, copier,
recording/answering machine,
scanner and printer
PHOTOGRAPHY
E. Chan

PRODUCT
General Electric Home Office
DESIGNER
Vent Design Associates
FIRM
Vent Design Associates
DESCRIPTION
A wall mounted computer system
containing power supply, multi CD
disk drive, xerographic printer, and
CPU board. The compact, yet open
architecture system allows easy
upgrades or exchange of units.
CLIENT
General Electric Plastics, Advance
Design & Development as part of
the GE 'Living Environments' study
PHOTOGRAPHY
Kelly O'Connor

PRODUCT
Financial Teller Assist Terminal
DESIGNER
Rian Dittmer
MANUFACTURER
NCR Corp.
CLIENT
NCR Corp.
PHOTOGRAPHY
AGI Photography

PRODUCT
Financial Teller Assist Terminal
DESIGNER
Paul Hemsworth
MANUFACTURER
NCR Corp.
CLIENT
NCR Corp.
PHOTOGRAPHY
AGI Photography

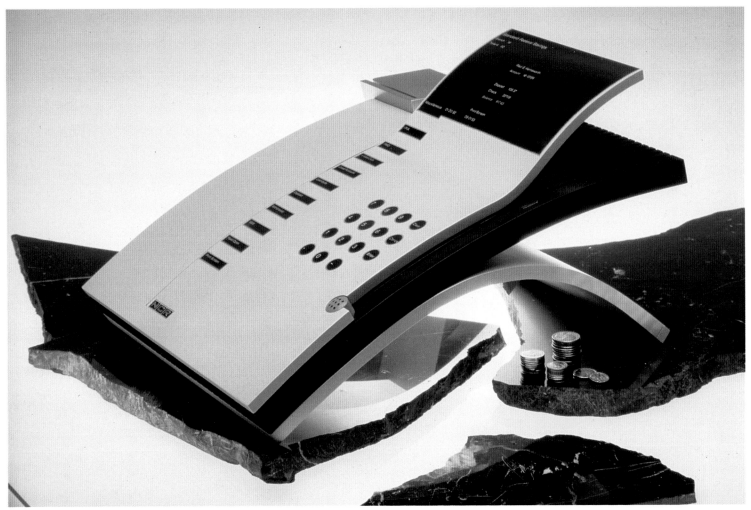

PRODUCT
IDB 2000 — Technical Reference
System
DESIGNER
Nelson Au
FIRM
Matrix Product Design
MANUFACTURER
Bell and Howell
DESCRIPTION
Image database for auto dealers;
monitor/display to replace
microfiche installations
CLIENT
Bell and Howell
PHOTOGRAPHY
Rick English

PRODUCT
Slate
DESIGNERS
Joel Carpenter, Tom Burchard,
Michael Dann
FIRM
The Design Works
MANUFACTURER
Telerate Systems, Inc.
DESCRIPTION
LCD touch screen provides data
to Wall Street traders
CLIENT
Telerate Systems, Inc.
PHOTOGRAPHY
Michael LaRiche

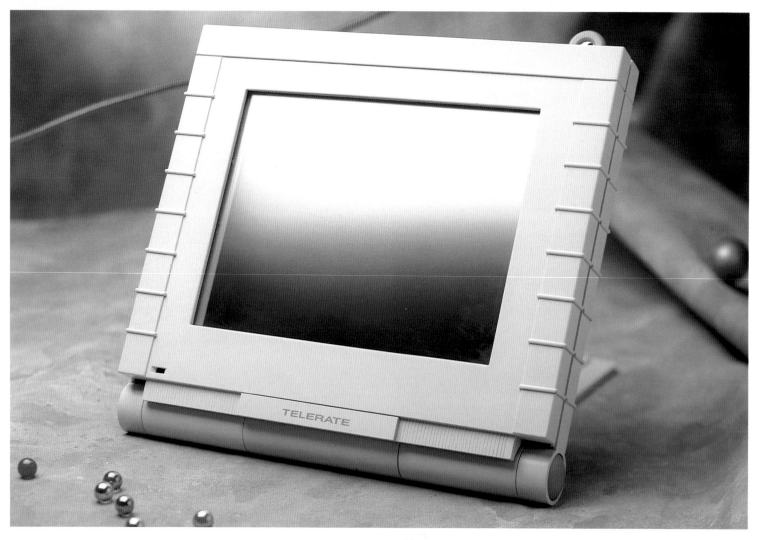

PRODUCT
Conceptual PC/Server
DESIGNER
Brian Jablonski
MANUFACTURER
NCR Corp.
CLIENT
NCR Corp.
PHOTOGRAPHY
AGI Photography

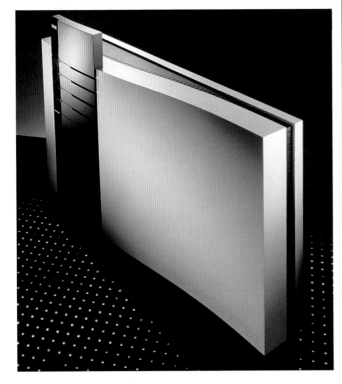

PRODUCT
Conceptual PC/Server
DESIGNER
Chet Wisniewski
MANUFACTURER
NCR Corp.
CLIENT
NCR Corp.
PHOTOGRAPHY
AGI Photography

PRODUCT
Outbound Laptop Computer
DESIGNER
Nelson Au
FIRM
Matrix Product Design
MANUFACTURER
Outbound Systems, Inc.
DESCRIPTION
Portable PC integrates
CLIENT
Outbound Systems, Inc.
PHOTOGRAPHY
Rick English

PRODUCT
Emily Julie & Max
DESIGNERS
Loyd Moore, Jeff Brown
FIRM
Technology Design
MANUFACTURER
Design Research
DESCRIPTION
Computer monitor, keypad and
CPU
CLIENT
Synapse
PHOTOGRAPHY
Jeff Curtis Photography

PRODUCT
Apple Computer, Inc. World
Keyboard
DESIGNER
Vent Design Associates
FIRM
Vent Design Associates
MANUFACTURER
Apple Computer, Inc.
DESCRIPTION
Apple's new standard issue full-
function height-adjustable
keyboard
CLIENT
General Electric Plastics, Advance
Design & Development as part of
the G.E. 'Living Environments'
study
PHOTOGRAPHY
Kelly O'Connor

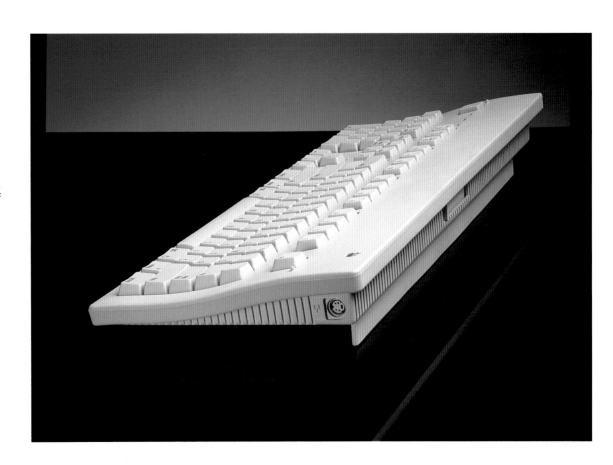

PRODUCT
Model 2207
DESIGNERS
Tom Burchard, Michael Dann
FIRM
The Design Works
MANUFACTURER
Numonics Inc.
DESCRIPTION
Low profile digitalizing tablet for
the P.C. CAD marketplace
CLIENT
Numonics Inc.
PHOTOGRAPHY
Michael LaRiche

PRODUCT
Piglet Personal Computer
DESIGNER
Geoff Hollington
FIRM
Hollington Associates
MANUFACTURER
Herman Miller Inc. USA
CLIENT
Herman Miller Inc. USA
PHOTOGRAPHY
Piers Bizony

PRODUCT
Dynabook Computer
DESIGNER
Mike Nuttall
DESIGN FIRM
Matrix Product Design, Inc.
MANUFACTURER
Dynabook Technology
DESCRIPTION
Portable PC
CLIENT
Dynabook Technology
PHOTOGRAPHY
Rick English

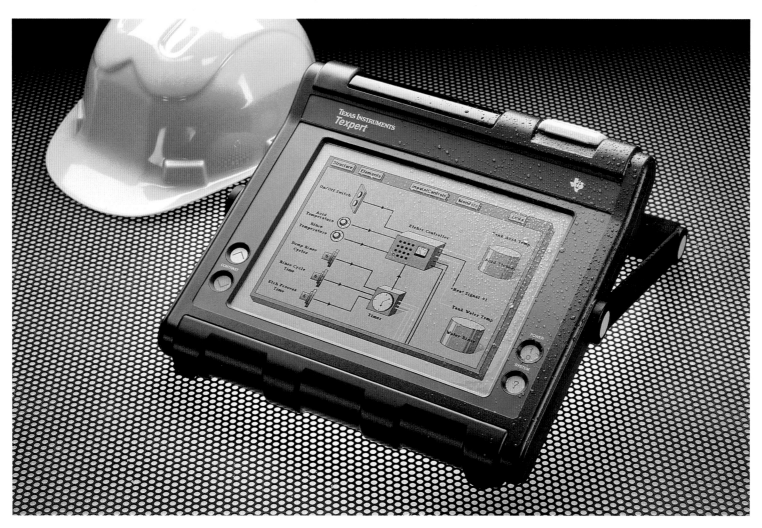

PRODUCT
Texpert
DESIGNER
Russell Stilley
MANUFACTURER
Texas Instruments
DESCRIPTION
Portable information system
CLIENT
Texas Instruments
PHOTOGRAPHY
Joseph Savant

PRODUCT
Yokozuna: Advanced Laptop System
DESIGNER
Khodi Feiz
FIRM
Texas Instrument Corporate
Design Center
DESCRIPTION
386/486 laptop system proposal;
emphasis is on integration of
system parts and the correct
chunking of components.
CLIENT
Texas Instruments

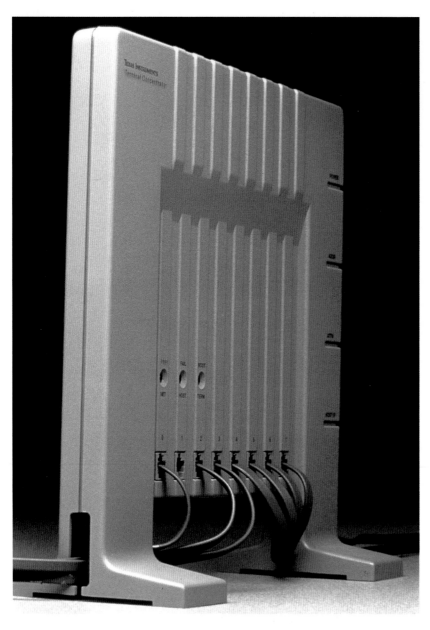

PRODUCT
Terminal Concentrator
DESIGNER
Khodi Feiz
FIRM
Texas Instrument Corporate
Design Center
DESCRIPTION
Concentrates up to eight terminal
cables into one unit, effectively
reducing the number of cables
going into the mainframe
CLIENT
Texas Instruments

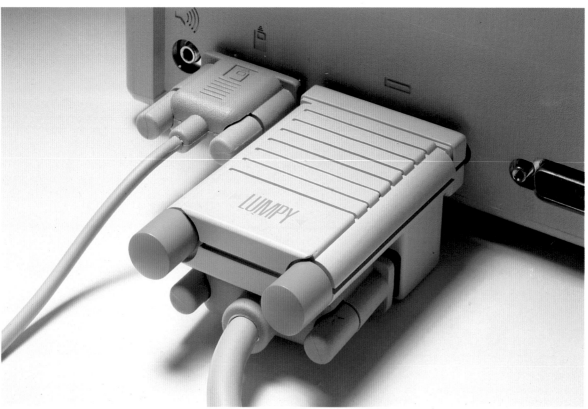

PRODUCT
MAC to DOS File Converter and
Drive
DESIGNER
Jeff Smith
FIRM
Lunar Design Inc.
MANUFACTURER
Kennect Tech.
CLIENT
Kennecht Tech.
PHOTOGRAPHY
Rick English

PRODUCT
Voice I.D. Unit (design study)
DESIGNER
Neumeister Design
FIRM
Neumeister Design
CLIENT
MBB Electronic Division

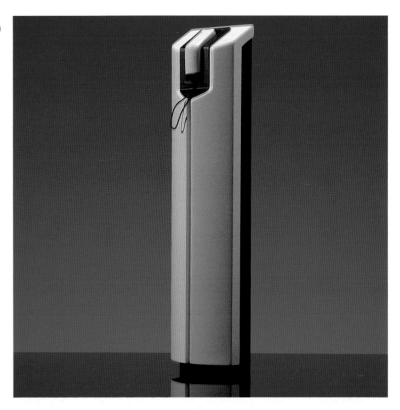

PRODUCT
PLA (Portable Logic Analyzer)
DESIGNER
Neumeister Design
FIRM
Neumeister Design
MANUFACTURER
Kontron Electronics
AWARD
1990 Design Center Stüttgart
Award

(detail)

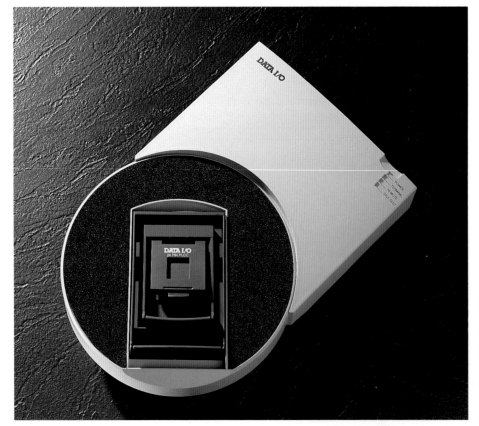

PRODUCT
Mesa I
DESIGNERS
Loyd Moore, Nick Barker
FIRM
Technology Design
MANUFACTURER
Data I/O
DESCRIPTION
In-circuit emulation pod for microprocessors
CLIENT
Data I/O

PRODUCT
2900 Programmer
DESIGNER
Loyd Moore, Jeff Brown, ED Eng.
FIRM
Technology Design
DESCRIPTION
Programs up to 24 different microprocessor chips
CLIENT
Data I/O
PHOTOGRAPHY
Jeff Curtis Photography

PRODUCT
TV Receiver
DESIGNER
Neumeister Design
FIRM
Neumeister Design
CLIENT
Loewe Opta GmbH

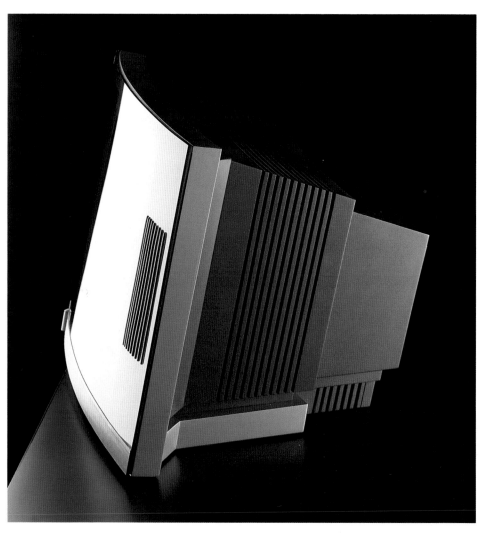

PRODUCT
IR — Telecommand Unit
(design study)
DESIGNER
Neumeister Design
FIRM
Neumeister Design
CLIENT
Loewe Opta GmbH

PRODUCT
ComGraph
DESIGNER
Neumeister Design
FIRM
Neumeister Design
MANUFACTURER
Messerschmitt Bölkow Blohm
DESCRIPTION
Graphic digitalizing work station

PRODUCT
NCube 2
DESIGNERS
Nick Barker, Jeff Brown, Mark Eike
FIRM
Technology Design
DESCRIPTION
Supercomputer based on hyper-
cube architecture; said to be the
fastest computer in the world.
CLIENT
NCube

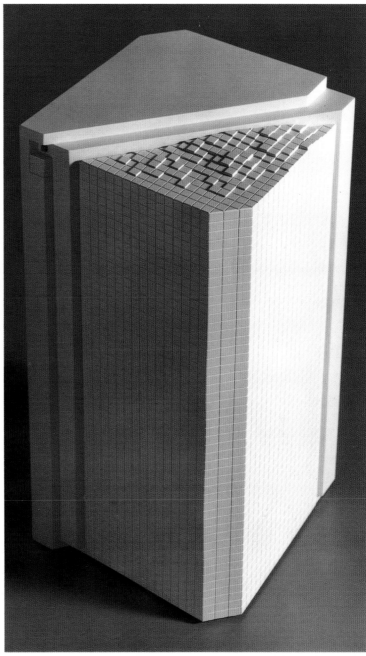

PRODUCT
Parallel Processing Computer
(current design)
DESIGNERS
Loyd Moore, Nick Barker,
Steve Kaneko
FIRM
Technology Design
DESCRIPTION
Enclosure for a near-super
computer
CLIENT
Floating Point Systems

PRODUCT
Parallel Processing Computer
(future design)
DESIGNERS
Loyd Moore, Nick Barker,
Steve Kaneko
FIRM
Technology Design
DESCRIPTION
Enclosure for a near-super
computer
CLIENT
Floating Point Systems

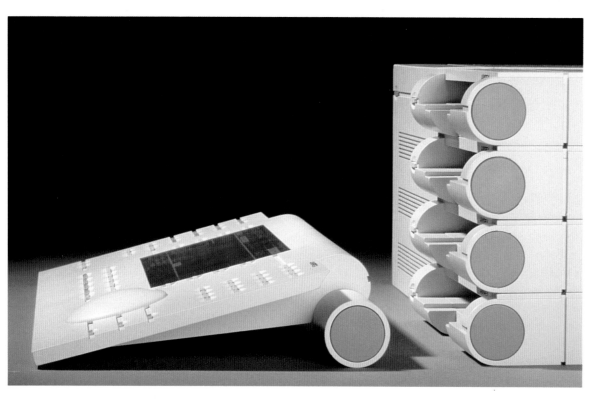

PRODUCT
Modular Professional 8mm Video
Recorder System
DESIGNER
Ron Boeder IDSA
FIRM
Ron Boeder Design
MANUFACTURER
Ampex Corp.
DESCRIPTION
Modular design allows unit to fit
under desk, tabletop or rack
mounting and system expansion
for production cutting
CLIENT
Ampex Corp.
PHOTOGRAPHY
Hudson Edwards
AWARD
IDEA Special Mention 1989

PRODUCT
MiniCom
DESIGNER
Neumeister Design
FIRM
Neumeister Design
MANUFACTURER
Messerschmitt Bölkow Blohm
DESCRIPTION
Personal video-conference tower

PRODUCT
MiniCom (detail)

PRODUCT
Video Com
DESIGNER
Neumeister Design
FIRM
Neumeister Design
MANUFACTURER
Messerschmitt Bölkow Blohm,
Space Division
DESCRIPTION
Video conference unit
AWARD
1989 Design Center Stüttgart
Award

PRODUCT
Intus 2000 Machine Data Terminal
DESIGNER
Neumeister Design
FIRM
Neumeister Design
MANUFACTURER
Periphere Computer Systeme
(PCS)
AWARD
1989 and 1990 "iF" Award for
Good Industrial Design

PRODUCT
Clover
DESIGNER
Mike Nuttall
FIRM
Matrix Product Design
MANUFACTURER
Silicon Graphics
DESCRIPTION
High end graphics workstation
CLIENT
Silicon Graphics
PHOTOGRAPHY
Rick English

PRODUCT
Chartviewer
DESIGNERS
Mark Eike, Steve Kaneko
FIRM
Technology Design
DESCRIPTION
Navigation map viewer
CLIENT
In Focus Systems
PHOTOGRAPHY
Jeff Curtis Photography

PRODUCT
SQUIN
DESIGNER
Ninaber/Peters/Krouwel
FIRM
Ninaber/Peters/Krouwel
MANUFACTURER
Holec
PHOTOGRAPHY
Ninaber/Peters/Krouwel
photostudio
AWARD
"ioN" Award for good industrial
design

197

PRODUCT
FISSO Desktop Set
DESIGNER
Takenobu Igarashi
FIRM
Igarashi Studio
DESCRIPTION
Set consisting of pen set, scissors set, postal scale, stamp pad, utility case, letter tray and display package
CLIENT
Raymay Fujii Corp.
PHOTOGRAPHY
Masaru Mera

PRODUCT
Jumbo Pen
FIRM
Canetti Design Group
MANUFACTURER
Kutsuwa
DESCRIPTION
Anodized aluminum ballpen
CLIENT
Canetti Inc.
PHOTOGRAPHY
Color Track

PRODUCT
Lunchboxes
DESIGNER
Nurit Amdur
FIRM
Panline USA, Inc.
MANUFACTURER
Panline USA, Inc.
CLIENT
Alex

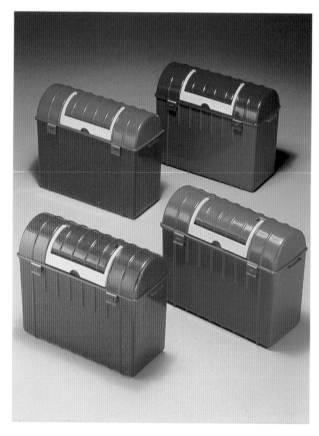

198

PRODUCT
Focux 90
DESIGNERS
Ravi Sawhney, Daniel Koo,
Scott Yu
FIRM
RKS Design
MANUFACTURER
Ellitech
DESCRIPTION
Hand-held laser pointer
PHOTOGRAPHY
Dan Hutchings, Ichiro Iwasaki,
Donald Brown
AWARD
Hanover Fair Design Selection

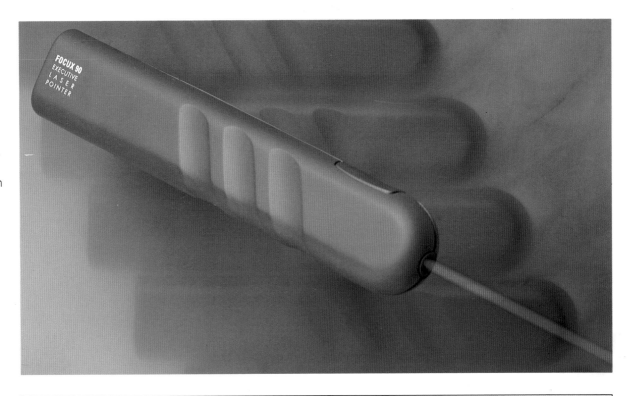

PRODUCT
"Escargot" Letter Opener and
Desktop Base
DESIGNERS
Gregory Hicks; Jeff Hanna
MANUFACTURER
Pat Crawford, Crawford Knives
DESCRIPTION
Ergonomically correct and
aesthetically pleasing letter
opener which requires only
minimal grip strength for
effective use
CLIENT
URO Designs
PHOTOGRAPHY
Bill Stumpf, Circle Studios

PRODUCT
Class 300 Tape Dispenser
DESIGNER
Marlan Polhemus
FIRM
Goldsmith Yamasaki Specht
MANUFACTURER
Tenex
DESCRIPTION
Unique "flat" profile tape
dispenser fits in drawer or on
desk
CLIENT
Tenex Corp.
PHOTOGRAPHY
Cabanban

PRODUCT
Scambio 9
DESIGNER
Massimo Iosa-Ghini
MANUFACTURER
Palazzetti

PRODUCT
Hollington Chair (Workchair)
DESIGNER
Geoff Hollington
FIRM
Hollington Associates
MANUFACTURER
Herman Miller Inc. USA
CLIENT
Herman Miller Inc. USA

PRODUCT
Canasta Chair/Executive Lowback
DESIGNER
Paolo Parigi
MANUFACTURER
Palazzetti

PRODUCT
Abak
DESIGNER
Studio Kairos
MANUFACTURER
B & B Italia
DESCRIPTION
Office Partitions
PHOTOGRAPHY
Luciano Soave

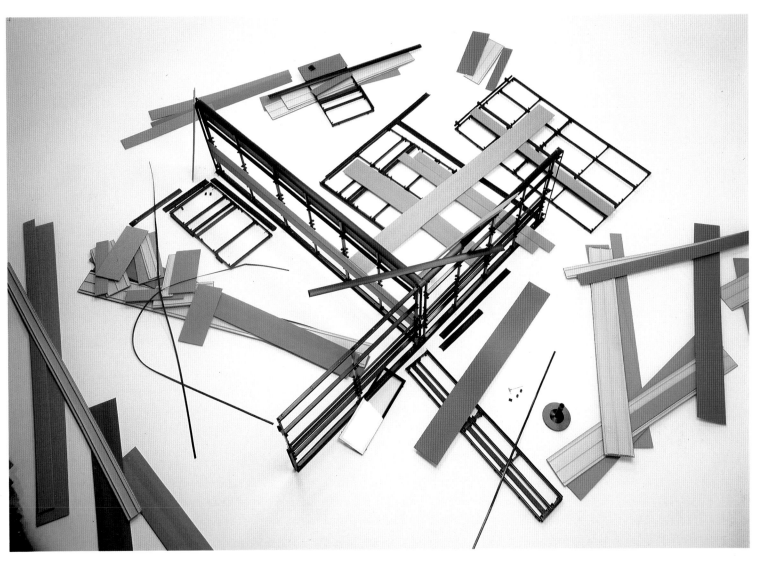

CHAPTER

8 Miscellaneous

DESIGN IS WHAT SEPARATES MAN from animals; it is the opposite of Fate. From the beginning, designers have been creating artifacts ranging from the functional to the artistic. People use those products not only to make their lives more comfortable, convenient and safe, but to satisfy inner desires, express personal ideas and communicate social mores. Just as "you are what you eat," we humans define ourselves by what we wear, where we live, our hairdos and

cars. Adrian Forty wrote in *Objects of Desire* that "design has the capacity to cast myths into enduring, solid and tangible form, so that they seem to be reality itself."

Design is exciting work because it is complex and dangerous. In modern times, utopian dreamers and capitalists sold us a future of labor saving gadgets. Designers have been in the business of fulfilling those dreams. But in 1990 it looks like we are traveling down the wrong road, because many of those gizmos and technologies are causing bigger problems. The environment is being destroyed, begging is outlawed, hunger is allowed, escapes and other drugs permeate every level of society. We now seem trapped by our own products.

In 1985, five years after Tucker Viemeister, Tom Dair and Tamara Thomsen, began working with Davin Stowell, they founded Smart Design Inc. Each of the four partners specialize in different aspects of design. From industrial design and development of consumer, medical and business products, to packaging, graphics and interface design, Smart Design is committed to the total design process.

The multi-faceted team of 17 is drawn from various disciplines; through the collaboration of research and imagination they expand the boundaries of what is possible. Mr. Viemeister is an enthusiastic advocate of the untapped power of design, not only for industry or individuals but to make a better world for all. Smart Design's work has been honored with many awards, is represented in the collections of Cooper-Hewitt and the Kunstergrasse Museums, is sold in the Museum of Modern Art, and has made lots of regular people happy.

TUCKER VIEMEISTER
Partner
SMART DESIGN INC.

PRODUCT
Crown GPW Walkie Pellet Truck
DESIGNERS
Gregory S. Breiding, David B. Smith
FIRM
Fitch RichardsonSmith
MANUFACTURER
Crown Equipment Co.

DESCRIPTION
Compact, battery-powered fork lift
CLIENT
Crown Equipment Co.
PHOTOGRAPHY
Courtesy of Crown Equipment Co.
AWARD
1989 IDSA/IDEA Honorable Mention

PRODUCT
Crown Turret Stockpicker
DESIGNERS
Gregory S. Breiding, David B.
Smith, Keith Kresge, Rainer Teufel
FIRM
Fitch RichardsonSmith
MANUFACTURER
Crown Equipment Co.

CLIENT
Crown Equipment Co.
PHOTOGRAPHY
Courtesy of Crown Equipment Co.
AWARDS
1989 Hanover Fair "if" Award;
1989 ID Annual Design Review;
1989 IDSA/IDEA First Place
Winner, Machinery Category

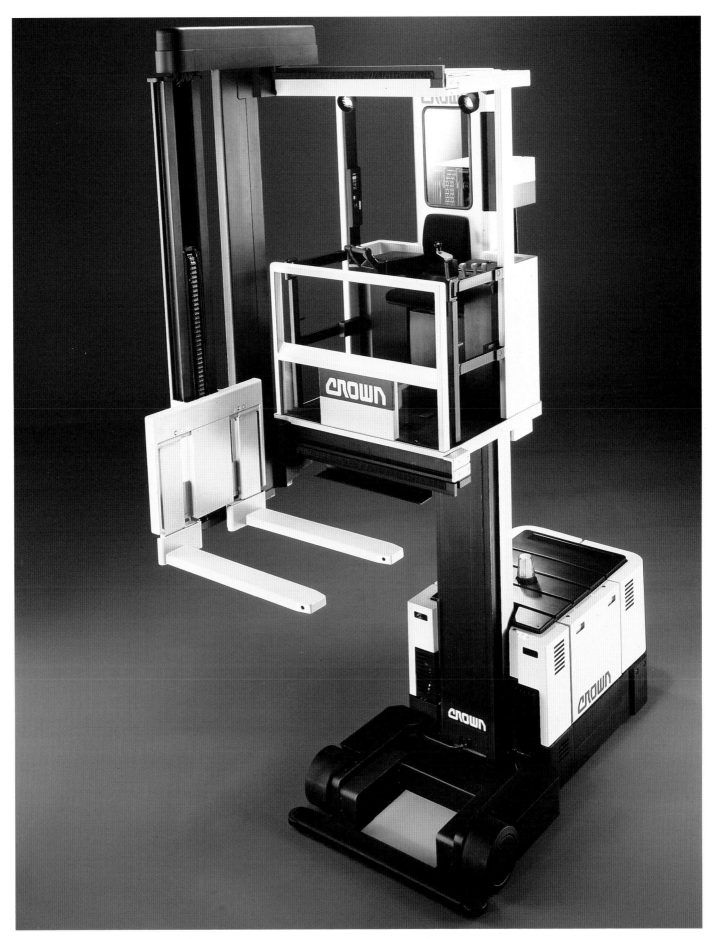

PRODUCT
Totem-Obelisk
DESIGNERS
Roberto Marcatti, Riccardo
Marcatti, Alfonso Crotti,
Gigi Conti
MANUFACTURER
Cooperativa Ceramica D'Imola
DESCRIPTION
Ceramic and iron obelisk
PHOTOGRAPHY
Foto Mosna

PRODUCT
Satellite Antenna Concept 2
(front view)
DESIGNERS
Jim Pagella, Charles Keane, Steven
Monti
FIRM
Gregory Fossella Design, Inc., a
division of the McCalla/Lackey
Corp.
DESCRIPTION
Size has been drastically reduced
with newly-developed technology
CLIENT
G.E. American Communications,
G.E. Co.
PHOTOGRAPHY
David Shopper

PRODUCT
Satellite Antenna Concept 1
(front view)
DESIGNERS
Jim Pagella, Charles Keane, Steven
Monti
FIRM
Gregory Fossella Design, Inc., a
division of the McCalla/Lackey
Corp.
DESCRIPTION
Size has been drastically reduced
with newly-developed technology
CLIENT
G.E. American Communications,
G.E. Co.
PHOTOGRAPHY
David Shopper

PRODUCT
Satellite Antenna Concept 1
(side view)

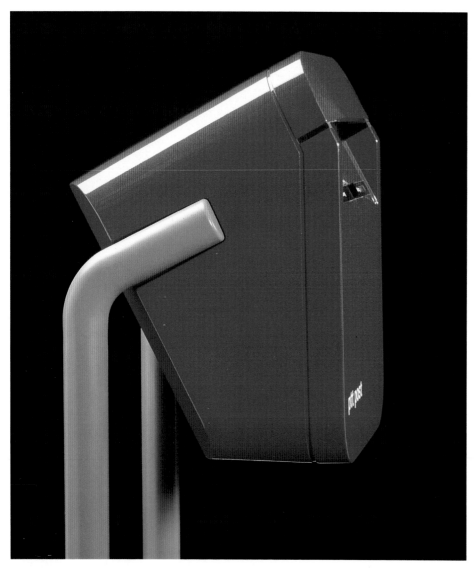

PRODUCT
Single mailbox
DESIGNER
Ninaber/Peters/Krouwel
FIRM
Ninaber/Peters/Krouwel
MANUFACTURER
Wientjes Polyester
CLIENT
Koninklijke PTT Nederland
PHOTOGRAPHY
Ninaber/Peters/Krouwel
photostudio

PRODUCT
Light Pole
DESIGNER
Gordon Randall Perry
FIRM
Gordon Randall Perry Design Inc.
MANUFACTURER
Cooper Lighting
DESCRIPTION
High-strength free-form
configured post made from
seamless aluminum tubing
CLIENT
Pfaff & Kendall (original
manufacturer)

PRODUCT
Traffic Barrier
DESIGNERS
Jeff Bransky, Thomas Purdue,
Leroy Goff
FIRM
Contours Design
DESCRIPTION
Designed to fold flat on impact
to reduce potential road hazard
of flying barriers
CLIENT
Warning Lights of Illinois
PHOTOGRAPHY
Jeff Bransky

PRODUCT
New York Newstand
DESIGNERS
Anderson/Schwartz Architects
and Calori & Vanden Eynden
FIRM
Anderson/Schwartz Architects
MANUFACTURER
ASAP
DESCRIPTION
Prototypical newstand for New
York City composed of the
elements of a printing press
PHOTOGRAPHY
Elliott Kaufman

PRODUCT
NEWSVENDER Newspaper Vending
Machine
DESIGNERS
Industrial & Mechanical Design:
Paul Arato, F. William Coffman
and Ken Cummings
Electronic Design: Willy Jidkoff
FIRM
Arato Designs Associates Inc.
MANUFACTURER
Kantrail Plastics Ltd.
DESCRIPTION
The top-opening feature includes
a display window for the paper,
positioned to be easily read by
people walking or driving past;
this also allows wheelchair users
easier access than from a tradi-
tional front-loader.
CLIENT
Kantrail Plastics Ltd.
PHOTOGRAPHY
Tony Lelkes

PRODUCT
InterCity Experimental
(in environment)
DESIGNER
Neumeister Design
FIRM
Neumeister Design
MANUFACTURER
Messerschmitt Bölkow Blohm,
Transport Division
DESCRIPTION
High-speed experimental train
CLIENT
German Federal Railways
AWARD
1987 Brunel Award

(interior)

(standard passenger section)

PRODUCT
Transrapid 07 "Europe"
(frontal view)
DESIGNER
Neumeister Design
FIRM
Neumeister Design
MANUFACTURER
Thyssen Henschel, New Transportation Systems Division
DESCRIPTION
Magnetic levitation high-speed train
AWARD
1990 Design Center Stüttgart Award

(interior)

PRODUCT
JCI Optima Energy Management System
DESIGNERS
Joe McArdle; Joe Munsch
FIRM
GVO, Inc.
DESCRIPTION
Used in high-rise building energy management of heat and airconditioning; constructed of structural foam (main door), diecast magnesium, injection-molded polycarbonate and sheet metal.
CLIENT
Johnson Controls
PHOTOGRAPHY
Mark Gottlieb, JCI Corporate Photography

PRODUCT
Texas Instruments PAC 200
DESIGNER
Khodi Feiz
FIRM
Texas Instrument Corporate Design Center
DESCRIPTION
The PAC 200 is a semi-conductor processing machine for the manufacture of silicon slices
CLIENT
Texas Instruments, Process Automation Center

PRODUCT
Sta-Temp Desoldering System
DESIGNER
Michael Barry
FIRM
GVO, Inc.
DESCRIPTION
Lightweight, ergonomic die-cast aluminum workstation for non-destructive repair of printed circuit boards
CLIENT
Metcal
PHOTOGRAPHY
Mark Gottlieb

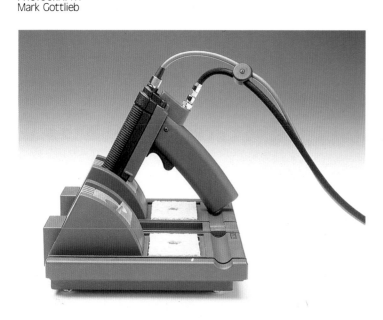

PRODUCT
Control Systems Test
Management Station (full view)
DESIGNER
Gregory Kenny; Cia Mooney
FIRM
Gregory Fossella Design,Inc., a
division of the McCalla/Lackey
Corp.
MANUFACTURER
McCalla/Lackey Productions
DESCRIPTION
Multi-use workstation for
generating test conditions and
monitoring the performance of
aircraft control systems
CLIENT
Grumman Aircraft Systems
PHOTOGRAPHY
Client Photograhic Staff

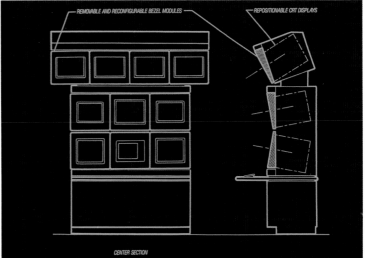

PRODUCT
Control Systems Test
Management Station
(computer image)

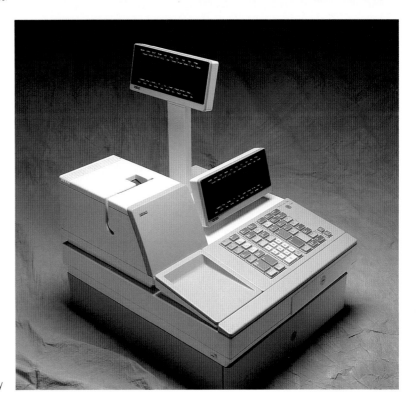

PRODUCT
NCR 2127 Electronic Cash
Register
DESIGNERS
S. Kato, Y. Nakamura
CLIENT
NCR Corp.
PHOTOGRAPHY
NCR Staff and AGI Photography

PRODUCT
Nuvola Ashtray
DESIGNER
Roberto Marcatti
FIRM
Lavori in Corso
MANUFACTURER
Lavori in Corso
DESCRIPTION
Aluminum floor ashtray with
ceramic inner element
PHOTOGRAPHY
Foto Mosna

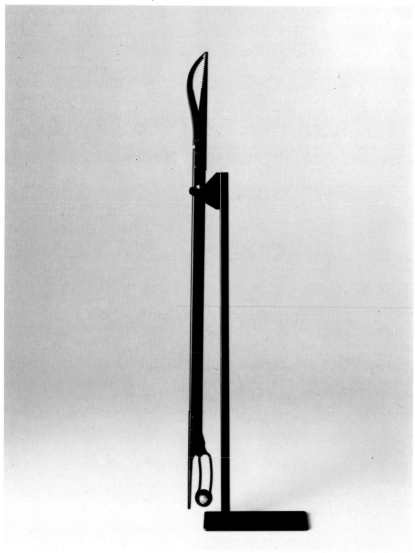

PRODUCT
Fireplace Tongs
DESIGNER
Pep Bonet
MANUFACTURER
Alessi
CLIENT
Markuse Corp.

PRODUCT
Brill
DESIGNER
Roberto Marcatti
FIRM
Mondial New Line
MANUFACTURER
Mondial New Line
DESCRIPTION
Brass and copper boutique set
includes height-adjustable shoe-
holder, shirt-holder, and hat
holder
PHOTOGRAPHY
Foto Mosna

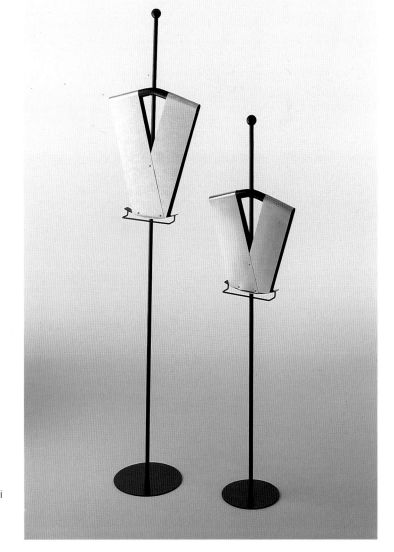

PRODUCT
Cavaliere
DESIGNER
Maurizio Peregalli
CLIENT
Noto
PHOTOGRAPHY
Bitetto-Chimenti

PRODUCT
Good Grips Gadgets
DESIGNERS
Davin Stowell, Tucker Viemeister,
Dan Formosa, Steve Russak,
Stephan Allendorf, Michael
Callahan
FIRM
Smart Design Inc.
MANUFACTURER
Oxo International Inc.
DESCRIPTION
Pizza wheel, scissors, can opener
CLIENT
Oxo International Inc.
PHOTOGRAPHY
Kenneth Willardt

PRODUCT
ExPress Closet Foldaway Ironing
Board (profile)
DESIGNERS
Scott Stropkay, Robert J. Hayes,
Lisa Stein
FIRM
Fitch RichardsonSmith
MANUFACTURER
Seymour Housewares Corp.
DESCRIPTION
Out of sight ironing surface that
glides down and rotates for flat
use in 3 seconds
CLIENT
Seymour Housewares Corp.
PHOTOGRAPHY
Steven Trank

PRODUCT
ExPress Closet Foldaway Ironing
Board (front view)

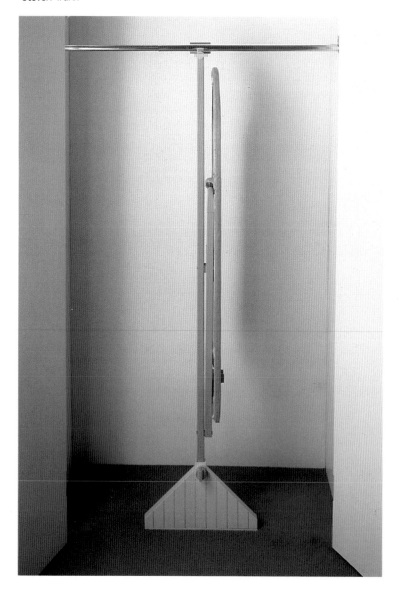

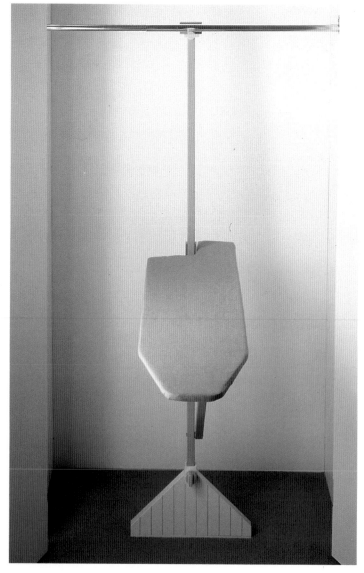

PRODUCT
Pocket Pocket
DESIGNER
Marianne Forrest
MANUFACTURER
Marianne Forrest
DESCRIPTION
Black oxidised silver and inlaid
18-carat gold, inspired by the
debris and dust found in the
average pocket
PHOTOGRAPHY
Marianne Forrest

PRODUCT
Matte Silver Watch
DESIGNERS
Olann Corp. and Eastern Accent
International Inc.
MANUFACTURER
Switzerland
PHOTOGRAPHY
Timothy Poosikian
DISTRIBUTOR
Eastern Accent International Inc.

PRODUCT
Pocket Watch
DESIGNER
Marianne Forrest
MANUFACTURER
Marianne Forrest
DESCRIPTION
Silver, 18-carat gold and monel
metal
PHOTOGRAPHY
Marianne Forrest

PRODUCT
Japanese Soaps
DESIGNER
Master Corp. Japan
FIRM
Eastern Accent International Inc.
DESCRIPTION
Five soap scents — gardenia, citron, mugwort, black sugar and rice bran — in ricepaper packaging
PHOTOGRAPHY
Julia Seltz
DISTRIBUTOR
Eastern Accent International Inc.

PRODUCT
Agricultural Sprayer
DESIGNER
Jaume Edo
FIRM
Via Design S.A.
DESCRIPTION
Shape adjusts to the anatomy of the user
CLIENT
Channel
PHOTOGRAPHY
Vandellos
AWARD
1987 Chamber of Commerce Award

PRODUCT
Visitor's Glasses
DESIGNER
Paul Specht
FIRM
Goldsmith Yamasaki Specht
MANUFACTURER
Encon
DESCRIPTION
High-impact plastic safety/ protection eyeglasses for commercial use
CLIENT
Encon
PHOTOGRAPHY
Cabanban

PRODUCT
Animal Workstuff
FIRM
Canetti Design Group
MANUFACTURER
Canetti Inc.
CLIENT
Canetti Inc.

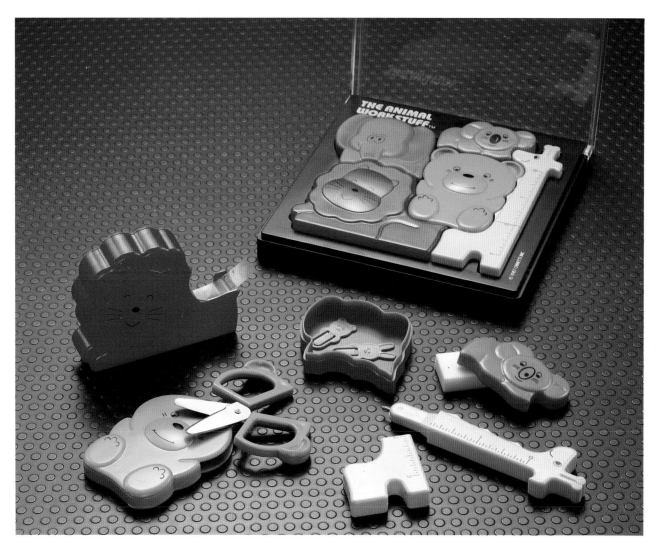

PRODUCT
Babybob
DESIGNER
Ninaber/Peters/Krouwel
FIRM
Ninaber/Peters/Krouwel
MANUFACTURER
Dremefa
DESCRIPTION
Cradle
PHOTOGRAPHY
Ninaber/Peters/Krouwel
photostudio

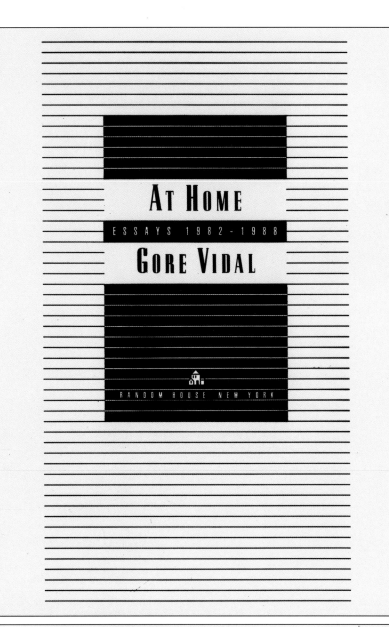

AT HOME

ESSAYS 1982-1988

GORE VIDAL

RANDOM HOUSE NEW YORK

PRODUCT
Title Page Design (At Home)
DESIGNER
James K. Lambert
CLIENT
Random House

14   At Home

public without worshipers—no other word—storming them. Yet each was obliged to spend a lot of time not only publicizing and selling aircraft but encouraging air transport. Of the two, Lindbergh was the better paid. But, as a deity, the commercial aspect was nothing to him, he claimed, and the religion all. On the other hand, Earhart's husband, the publisher and publicist George Palmer Putnam (known as G.P.), worked her very hard indeed. The icons of the air age were big business.

Time magazine, September 28, 1931:

To Charles Townsend Ludington, socialite of Philadelphia, $8,000 might be the price of a small cabin cruiser such as he sails on Biscayne Bay. . . . But the $8,073.61 profit which showed on a balance sheet upon [his] desk last week was as exciting to him as a great fortune. It was the first year's net earning of the Ludington Line, plane-per-hour passenger service between New York, Philadelphia and Washington.

As practically sole financiers of the company [Nicholas and Charles Townsend] Ludington might well be proud. But they would be the first to insist that all credit go to two young men who sold them the plan and then made it work: brawny, handsome Gene Vidal, West Point halfback of 1916–20, one-time Army flyer; and squint-eyed, leathery Paul ("Dog") Collins, war pilot, old-time airmail pilot.

*Time*style still exerts its old magic, while *Time*checkers are, as always, a bit off—my father graduated from West Point in 1918. An all-American halfback, he also played quarterback. But he *was* one of the first army flyers and the first instructor in aeronautics at West Point. Bored with peacetime army life and excited by aviation, he quit the army in 1926. Already married to the "beauteous" (*Time* epithet) Nina Gore, daughter of "blind solon" (ditto) Senator T. P. Gore, he had a year-old son for whom *Time* had yet to mint any of those Lucite epithets that, in time (where "All things shall come to pass," Ecclesiastes), they would.

≡

New airlines were cropping up all over the country. After 1918, anyone who could nail down a contract from the postmaster general to fly the mail was in business. Since this was the good old United States, there was corruption. Unkind gossips thought that an army flyer whose father-in-law was a senator would be well placed to get such a contract. But during the last years of President Hoover, Senator Gore was a Democrat; and during

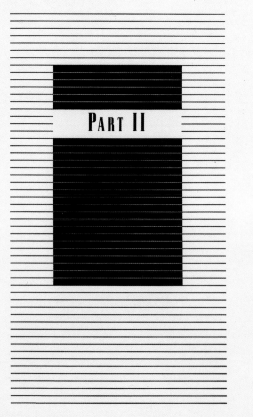

PART II

PRODUCT
Text Page and Part Title
(At Home)
DESIGNER
James K. Lambert
CLIENT
Random House

222

PRODUCT
Text Page and Chapter Opener
(At Home)
DESIGNER
James K. Lambert
CLIENT
Random House

tee to investigate the former Republican postmaster general Brown's dealings with the airlines. Black's highly partisan committee painted Brown even darker than he was. Yes, he had played favorites in awarding mail contracts but no one could prove that he—or the Grand Old Party—had in any specific way profited. Nevertheless, Jim Farley, the new postmaster general, charged Brown with "conspiracy and collusion" while the president, himself a man of truly superhuman vindictiveness, decided to punish Brown, the Republican party, and the colluded-with airlines.

What could be more punitive—and dramatic—than the cancellation of all U.S. airmail contracts with private companies? Since the army had flown the mail back in 1918, let them fly the mail now. The president consulted the director of Air Commerce, who told him that army flyers did not have the sort of skills needed to fly the mail. After all, he should know; he was one. Undeterred, the president turned to General Benjamin D. Foulois, the chief of the air corps, who lusted for appropriations as all air corps chiefs do; and the general said, of course, the air corps could fly the mail.

On February 9, 1934, by executive order, the president canceled all airmail contracts; and the Army flew the mail. At the end of the first week, five army pilots were dead, six critically injured, eight planes wrecked. One evening in mid-March, my father was called to the White House. As Gene pushed the president's wheelchair along the upstairs corridor, the president, his usual airy self, said, "Well, Brother Vidal, we seem to have a bit of a mess on our hands." Gene always said, "I found that 'we' pretty funny." But good soldiers covered up for their superiors. What, FDR wondered, should they do? Although my father had a deep and lifelong contempt for politicians in general ("They tell lies," he used to say with wonder, "even when they don't have to") and for Roosevelt's cheerful mendacities in particular, he did admire the president's resilience: "He was always ready to try something new. He was like a good athlete. Never worry about the last play. Only the next one." Unfortunately, before they could extricate the administration from the mess, Charles Lindbergh attacked the president; publicly, the Lone Eagle held FDR responsible for the dead and injured army pilots.

Roosevelt never forgave Lindbergh. "After that," said Gene, "he would always refer to Slim as 'this man Lindbergh,' in that condescending voice

PRODUCT
Title Page Design
DESIGNER
James K. Lambert
CLIENT
Random House

CHAPTER 4

TENNESSEE WILLIAMS: SOMEONE TO LAUGH AT THE SQUARES WITH

1

Although poetry is no longer much read by anyone in freedom's land, biographies of those American poets who took terrible risks not only with their talents but with their lives, are often quite popular; and testimonies, chockablock with pity, terror and awe, provide the unread poet, if not his poetry, with a degree of posthumous fame. Ever since Hart ("Man overboard!") Crane dove into the Caribbean and all our hearts, the most ambitious of our poets have often gone the suicide route:

There was an unnatural stillness in the kitchen which made her heart skip a beat; then she saw Marvin, huddled in front of the oven; then she screamed: the head of the "finest sestina-operator of the Seventies" [*Hudson Review*, Spring 1971] had been burned to a crisp.

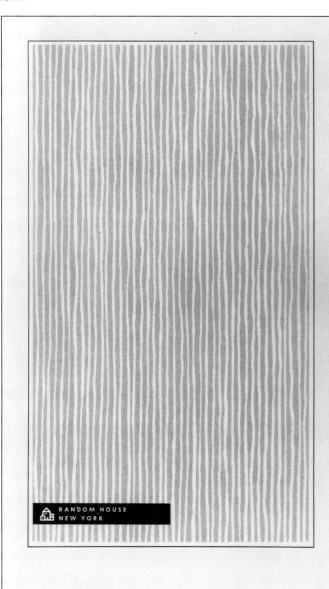

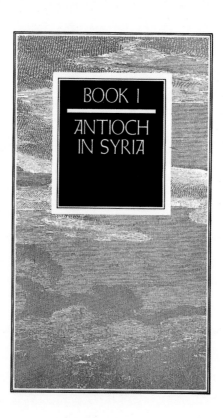

PRODUCT
Title Page (Soul Flame)
DESIGNER
James K. Lambert
CLIENT
Random House

Strangely, the infant boy, still cradled against its mother's breast, had not yet made a sound.

At long last Mera was able to guide the second baby onto the waiting sheets. With profound relief she saw that it lived. But as she was cutting the cord, she heard a sound outside mingle with the wind, a sound that should not have been there. Mera brought her head up sharply and saw that the Roman was staring at the door.

"Horses," he said. "Soldiers."

Then there was thunderous banging on the door, not of someone knocking to come in, but of someone trying to break it down.

"They have found us," he said simply.

Mera was up on her feet in an instant. "Come!" she hissed and ran for the narrow door at the end of the room. She did not look back, did not see the red-cloaked soldiers burst in; without thinking, she plunged into the darkness of the storage lean-to that abutted her house and, with the newborn baby girl clasped wet and naked to her breast, climbed into the corncrib, curling herself as small as she could under the husks. As she huddled in the dark night of the crib, her skin pricked by the corn, barely breathing, Mera listened to the stamp of hobnailed sandals on the hard-packed dirt floor. There was a brief dialogue in Greek, a staccato demand and a reply, the whistle of metal through the air. Two sharp cries and then: silence.

Mera shivered uncontrollably. The baby trembled in her arms. Heavy footfalls sounded around the room, with one foray into the lean-to. Through the cracks in the bin she saw a light: someone was searching with a lamp. And then she heard the voice of the handsome Roman, weak and breathy: "There is no one, I tell you. The midwife was not at home. We are alone. I . . . *I* delivered the child myself . . ."

To her horror the baby in her arms started to whimper. Mera quickly placed her hand over the little face and whispered, "Blessed Mother, Queen of Heaven, don't let this baby be killed."

She held her breath again and listened. Now there was nothing around her but darkness and silence and the moaning wind. She waited. With the baby pressed to her bosom, her hand over its mouth, Mera lay crouched in the corncrib for what seemed like hours. Her body began to ache, the baby squirmed. But still she remained in hiding.

Finally, after what seemed an eternity, Mera thought she heard another voice in the wind. "Woman . . . " it called.

Cautiously she raised up. In the gloom of pre-dawn, Mera could just make out a crumpled form on the floor of the room, and she heard the Roman call out weakly, "Woman, they have gone . . ."

PRODUCT
Text Page and Part Title Page
(Soul Flame)
DESIGNER
James K. Lambert
CLIENT
Random House

PRODUCT
a hundred LEGENDS
DESIGNER
Don Ruddy
DESCRIPTION
A collection of art, photographs, poetry, prose, music and other media created by 127 people with AIDS. It represents a memorial to those living through the AIDS crisis and serves as a historic document, honors the artists and connects creativity with healing.
CLIENT
Design Industries Foundation for AIDS (DIFFA)
DISTRIBUTOR
DIFFA

9 Fantasy

WE ARE THE PRODUCTS WE USE.

Such an extreme reduction seems antithetical to the thought processes that govern intelligence. The genesis of thought does not include the act of becoming. The act of becoming something is a metaphysical question. We are quite complex and our products are an outgrowth of our thoughts and experiences. We do not emulate the product. Rather, the product is an extension of us in the exact sense that a telescope, a microscope or a camera is an extension of the human eye. We do not become something by the mere use of it. In any case, products reflect the best and the worst of us.

My interest is primarily in the relationship between art and design. Good design should encompass art, form and function. Varying degrees of these basic qualities can be found in most objects, and my design philosophy is to try to tip the balance and emphasize the aesthetic side of product design.

I think it is clear that in the future, people will be concerned with objects that last and are ecologically sound. The era of "Throw Away" culture seems to be coming to an end, and with this new concern for the planet will come a focus on recycled, hand-crafted and found object art. Design will have to reflect an intelligent approach to the environment, as well as being soundly utilitarian. The inter-relationship between the planet and the products will become more kindred than ever. We will have to become far more conscious of our effect on each other, our resources and our planet.

LOIS LAMBERT
Director
GALLERY OF FUNCTIONAL ART

Lois Lambert grew up in Chicago and was heavily influenced by the art and architecture of that city. She majored in Fine Arts and Psychology, holds a degree in Communications Arts and spent several years at the Art Institute of Chicago.

Ms. Lambert has traveled extensively, working as an art consultant, an interior designer, and a curator of special exhibits of Art Furniture. She has also lectured to architects, designers and the public on the subject.

The Gallery of Functional Art, owned and operated by Ms. Labert, opened in September 1988 and has won the Metropolitan Home Design 100 Award, as well as an "Angie" from <u>Angeles</u> Magazine.

PRODUCT
Ceiling Fixture
DESIGNER
Philip Miller
DESCRIPTION
Made of steel, brass, copperplate
and slumped glass (glass by Paul
Fisher)
CLIENT
Jeff and Susan Bridges
PHOTOGRAPHY
Philip Miller

PRODUCT
Hermes Recycling Fountain
DESIGNER
Chris Collicott
MANUFACTURER
Chris Collicott
DESCRIPTION
Indoor/outdoor fountain of
copper and aluminum
PHOTOGRAPHY
Chris Collicott

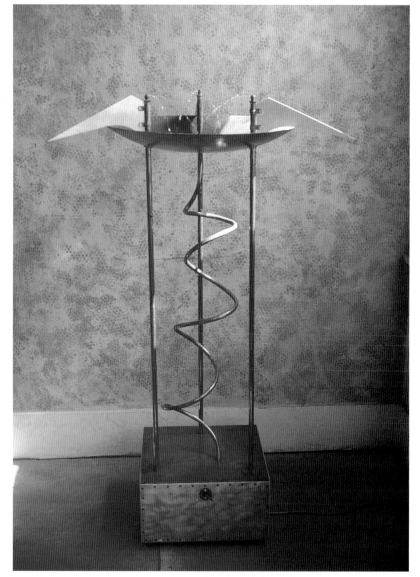

PRODUCT
"Waterfall" Clothing Rack
DESIGNER
Philip Miller
CLIENT
Companie BX
PHOTOGRAPHY
Philip Miller

228

PRODUCT
"Samoa" Hammock 1988
DESIGNER
Forrest Myers
MANUFACTURER
Forrest Myers and Art Et
Industrie

PRODUCT
Parlour Table
DESIGNER
Chris Collicott
MANUFACTURER
Chris Collicott
DESCRIPTION
Maple and steel table with a
velvet-lined drawer
PHOTOGRAPHY
Chris Collicott

PRODUCT
La Madonna Miro Chair
DESIGNER
Chris Collicott
MANUFACTURER
Chris Collicott
DESCRIPTION
Powder-coated steel chair with a
padded back and seat containing,
in the 'halo,' a working clock
PHOTOGRAPHY
Chris Collicott

229

PRODUCT
Chair
DESIGNER
Iris Fingerhut
PHOTOGRAPHY
Nicole Katano

PRODUCT
"The Crisis in Gourmet Marxism"
Dining Table
DESIGNER
David Gale
MANUFACTURER
David Gale
DESCRIPTION
Of steel tubing with cast-aluminum letters under a sheet of glass
CLIENT
Gallery of Functional Art
PHOTOGRAPHY
Tom Bonner

PRODUCT
Chair
DESIGNER
David Gale
MANUFACTURER
David Gale
DESCRIPTION
Steel-tube frame, cast-aluminum letters; a series of ten
CLIENT
Gallery of Functional Art

PRODUCT
Nightables
DESIGNER
Mark Robbins
MANUFACTURER
Mark Robbins
DESCRIPTION
A series of architectural studies
employing weights, mirrors and
double-hinged frames. Each piece
transforms revealing geometric
and figurative elements within
covered internal chambers.
PHOTOGRAPHY
Grant Taylor

PRODUCT
Ubangi Chair
DESIGNER
Philip Miller
DESCRIPTION
One of a series
PHOTOGRAPHY
Philip Miller

PRODUCT
Catable
DESIGNER
Dane Scarborough
MANUFACTURER
Dane Scarborough
DESCRIPTION
Coffee or end table of solid ash
and bending poplar
PHOTOGRAPHY
Landis McIntire
DISTRIBUTOR
Tim Wells Furniture

PRODUCT
Lance Lamp
DESIGNER
Fred Baier
FIRM
Fred Baier and Tim Wells
Partnership
MANUFACTURER
Tim Wells
DESCRIPTION
Stained maple
PHOTOGRAPHY
Cheryl Klauss
DISTRIBUTOR
Tim Wells Furniture

PRODUCT
Tiramisu Lamp
DESIGNER
Roberto Marcatti
FIRM
Noto-Zeus
MANUFACTURER
Noto-Zeus
DESCRIPTION
Steel floor lamp with metal
decoration
PHOTOGRAPHY
Foto Mosna

PRODUCT
Mei-mei's Lamp
DESIGNER
Richard Tuttle
FIRM
A/D
DESCRIPTION
Pine, 1890's molded glass shade
PHOTOGRAPHY
Ken Schles
DISTRIBUTOR
A/D

PRODUCT
Clock
DESIGNER
Philippe Starck
MANUFACTURER
Alessi
CLIENT
Markuse Corp.

PRODUCT
Heraldic Clock
DESIGNER
Chris Collicott
MANUFACTURER
Chris Collicott
DESCRIPTION
Wall clock made from bent steel
rod and kitchen utensils
PHOTOGRAPHY
Chris Collicott

PRODUCT
Phoenix Cuckoo Clock
DESIGNER
Chris Collicott
MANUFACTURER
Chris Collicott
DESCRIPTION
Copper and brass clock, a scarab
beetle 'cuckoo' pops out on the
hour and half-hour
PHOTOGRAPHY
Chris Collicott

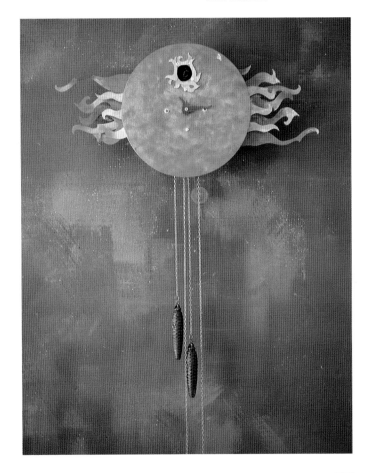

PRODUCT
Standing Clock
DESIGNER
Chris Collicott
MANUFACTURER
Chris Collicott
DESCRIPTION
Steel with goldleaf accents
PHOTOGRAPHY
Chris Collicott

PRODUCT
Time Table
DESIGNER
Chris Collicott
MANUFACTURER
Chris Collicott
DESCRIPTION
White oak table containing a
working clock under the glass top
PHOTOGRAPHY
Chris Collicott

PRODUCT
His and Hers Andirons
DESIGNER
Frederic Schwartz
FIRM
Anderson / Schwartz Architects
MANUFACTURER
Solesbury Forge
DESCRIPTION
Hand-forged black iron with brass
ball accents

234

PRODUCT
Water Kettle
DESIGNER
Philippe Starck
MANUFACTURER
Alessi
CLIENT
Markuse Corp.

FRONT VIEW

PRODUCT
Bollitore
DESIGNER
Philippe Starck
MANUFACTURER
Alessi
CLIENT
Markuse Corp.

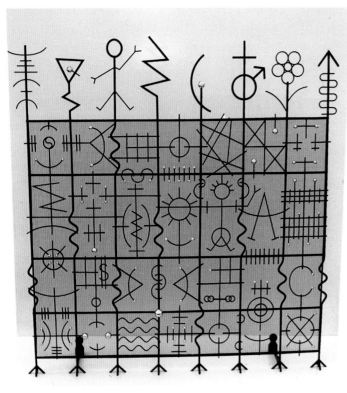

PRODUCT
Symbols of the Universe
Firescreen
DESIGNER
Frederic Schwartz
FIRM
Anderson / Schwartz Architects
MANUFACTURER
Solesbury Forge

PRODUCT
Citrus juicer
DESIGNER
Philippe Starck
MANUFACTURER
Alessi
CLIENT
Markuse Corp.

PRODUCT
Close Encounters of the Crystal
Kind Bolo Brooch
DESIGNER
Marc Cohen
FIRM
Still Life Creations
MANUFACTURER
Marc Cohen
DESCRIPTION
Wearable art (3-D miniatures)
PHOTOGRAPHY
Dominique Ragueneau

PRODUCT
Bolo Brooch
DESIGNER
Marc Cohen
FIRM
Still Life Creations
MANUFACTURER
Marc Cohen
DESCRIPTION
Wearable art (3-D miniatures)
PHOTOGRAPHY
Dominique Ragueneau

PRODUCT
Birdhouse
DESIGNER
Laura Foreman
CONCEPT BY
Laura Foreman; Kevin Sutton, technical assistance
DESCRIPTION
One of a kind slanted wood birdhouse sculpture with clapboard siding; room interior has a tableau with dilapidated furnishings. Hidden in the roof is a cassette player and headphones extending from the work enable viewer to hear selections from Charlie "Bird" Parker's "Ornithology." "I am as a sparrow alone on the rooftop."
PHOTOGRAPHY
Karen Bell
REPRESENTATIVE
Souyun Yi Gallery

PRODUCT
Tenement Birdhouse
DESIGNER
Laura Foreman
CONCEPT BY
Laura Foreman; John Esposito, technical assistance
DESCRIPTION
One of five birdhouse sculptures which are site-specific and functional, this particular work mirrors the tenement lived in by a beloved neighborhood resident, Mary Frances Carpenter, a paraplegic, and is in memorial of her. Work contains miniature air conditioners and fire escapes.
PHOTOGRAPHY
Karen Bell

PRODUCT
Guitar
DESIGNERS
Roger Pedlar
FIRM
FM Design
DESCRIPTION
The Index guitar, a remote
controller allowing guitar players
access to the sounds available
from any MIDI synthesizer.

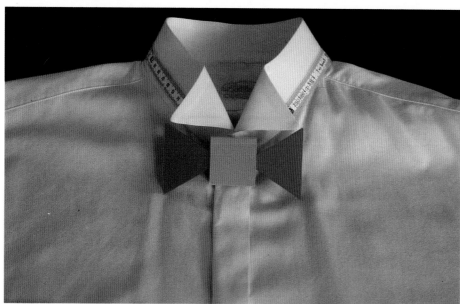

PRODUCT
Designer's Bow Tie
DESIGNER
Christoph Boeninger
DESCRIPTION
Doubles as a red and blue pencil
PHOTOGRAPHY
C. Boeninger

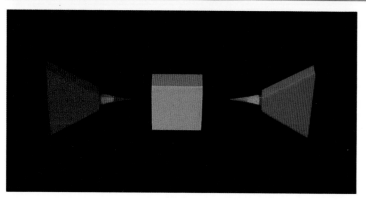

PRODUCT
Satori
DESIGNER
Peter Stathis
DESCRIPTION
Personal television with a hybrid
form that assumes a variety of
gestures/positions

(extended)

PRODUCT
Primitive Androyd TV Monitor
DESIGNER
Shozo Toyohisa
FIRM
Eastern Accent International Inc.
DISTRIBUTOR
Eastern Accent International Inc.

PRODUCT
"Shikimi, Nijo Daime-Shin"
DESIGNER
Shigeru Uchida
FIRM
Studio 80
MANUFACTURER
Chairs
DESCRIPTION
Easily-assembled and disassembled,
composed based on the two- and
three-fourth tatami floor space
PHOTOGRAPHY
Mitsumasa Fujitsuka

PRODUCT
"Shikimi, Nijo Daime-Gyo"
DESIGNER
Shigeru Uchida
FIRM
Studio 80
MANUFACTURER
Chairs
DESCRIPTION
Easily-assembled and disassembled,
composed based on the two- and
three-fourth tatami floor space

PRODUCT
"Shikimi, Nijo Daime-So"
DESIGNER
Shigeru Uchida
FIRM
Studio 80
MANUFACTURER
Chairs
DESCRIPTION
Easily-assembled and disassembled,
composed based on the two- and
three-fourth tatami floor space

241

Index